Lectures on Complex Approximation

Dieter Gaier

Lectures on Complex Approximation

Translated by Renate McLaughlin

Birkhäuser
Boston · Basel · Stuttgart

Dieter Gaier
Mathematisches Institut
Justus-Liebig-Universität
Arndtstr. 2
D-6300 Giessen
West Germany

Translator:
Renate McLaughlin
Mathematics Department
University of Michigan-Flint
Flint, MI 48502
USA

Library of Congress Cataloging in Publication Data
Gaier, Dieter.
Lectures on complex approximation.
Translation of: Approximation im Komplexen.
Bibliography: p.
Includes index.
1. Approximation theory. 2. Functions of complex variables. I. Title.
QA297.5.G3413 1985 515.9 83-3911

CIP-Kurztitelaufnahme der Deutschen Bibliothek
Gaier, Dieter:
Lectures on complex approximation / Dieter Gaier.
Transl. by Renate McLaughlin. — Boston ; Basel ;
Stuttgart : Birkhäuser, 1985.
Dt. Ansg. u.d.T.: Gaier, Dieter: Vorlesungen
über Approximation im Komplexen
ISBN 3-7643-3147-X (Boston)
ISBN 0-8176-3147-X (Basel)

This English edition is translated from *Vorlesungen über Approximation im Komplexen*, Birkhäuser Verlag, 1980

ISBN 3-7643-3147-X
ISBN 0-8176-3147-X

9 8 7 6 5 4 3 2 1
Printed in USA

TABLE OF CONTENTS

Preface to the English Edition xi
Preface to the German Edition xiii
Symbols and Notation xv

Part I: Approximation by Series Expansions and by Interpolation

Chapter I. Representation of complex functions by orthogonal series and Faber series 2

§1. The Hilbert space $L^2(G)$ 2
A. Definition of $L^2(G)$ 2
B. $L^2(G)$ as a Hilbert space 4

§2. Orthonormal systems of polynomials in $L^2(G)$ 6
A. Construction of ON systems; Gramian matrix 6
A_1. The Gram-Schmidt orthogonalization process 6
A_2. The Gramian matrix 7
A_3. A special case: Polynomials in $L^2(G)$ 9
B. Zeros of orthogonal polynomials 11
C. Asymptotic representation of the ON polynomials 12
Remark about §2 16

§3. Completeness of the polynomials in $L^2(G)$ 16
A. The problem and examples 16
B. Domains with the PA property 17
C. Domains not having the PA property 20
C_1. Slit domains 20
C_2. Moon-shaped domains 20
Remarks about §3 23

§4. Expansion with respect to ON systems in $L^2(G)$ 24
A. ON expansions in Hilbert space 24
B. ON expansions in the space $L^2(G)$ 25

C. The quality of the approximation if f is analytic in $\overline{G}$ 26
Remarks about §4 29

§5. The Bergman kernel function 30
A. Introduction and properties of the kernel function 30
B. Series representation of the Bergman kernel function 31
C. Construction of conformal mappings with the Bergman kernel function 32
C_1. The connection between K and conformal mapping 33
C_2. The Bieberbach polynomials 34
C_3. The use of singular functions in the ON process 36
D. Additional applications of the Bergman kernel function 37
D_1. Domains with the mean-value property 37
D_2. Representation of $\int_{-1}^{+1} f(x)dx$ as an area integral 37
Remark about §5 40

§6. The quality of the approximation; Faber expansions 40
A. Boundary behavior of Cauchy integrals 41
B. Faber polynomials and Faber expansions 42
C. The Faber mapping as a bounded operator 45
C_1. Curves of bounded rotation 45
C_2. The Faber mapping T 47
D. The quality of approximation inside a curve of bounded rotation 51
D_1. Preparations; uniform convergence 51
D_2. The modulus of continuity of the Cauchy integral corresponding to h 52
D_3. The quality of the approximation 53
E. Report on additional results 55
E_1. Additional uniform estimates 55
E_2. Local estimates 56
Remarks about §6 57

Chapter II. Approximation by interpolation 58

§1. Hermite's interpolation formula 58
A. The interpolating polynomial 58
B. Special cases of Hermite's formula 59

§2. Interpolation in uniformly distributed points; Fejér points, Fekete points 62
A. Preparations; rough statement about convergence 62
B. General convergence theorem of Kalmár and Walsh 64
C. The system of Fejér points 67
D. The system of Fekete points 70
Remarks about §2 72

§3. Approximation on more general compact sets; Runge's theorem 73
A. Again: Interpolation in Fekete points 73
B. Runge's approximation theorem 76
Remark about §3 78

§4. Interpolation in the unit disk 78
A. Interpolation on $\{z: |z| = r\}, r < 1$ 78
B. Interpolation on $\{z: |z| = 1\}$ 81
C. Approximation by rational functions 86
Remarks about §4 87

Part II: General Approximation Theorems in the Complex Plane

Chapter III. Approximation on compact sets 92

§1. Runge's approximation theorem 92
A. General Cauchy formula 93
B. Runge's theorem 94
C. The pole shifting method 95

§2. Mergelyan's theorem 97
A. Formulation of the result; special cases; consequences 97
B. Preparations for the proof 98
B_1. Tietze's extension theorem 99
B_2. A representation formula 99
B_3. Koebe's ¼-theorem 100
B_4. Mergelyan's lemma 101
C. Proof of Mergelyan's theorem 104
Remark about §2 109

§3. Approximation by rational functions 109
A. Swiss cheese 110
A_1. Alice Roth's construction 110
A_2. Swiss cheese with interior points 112

A_3. Swiss cheese with two components 113
A_4. Accumulation of holes at the diameter of $\mathbb{D}$ 113
B. Preparations for Bishop's theorem 114
B_1. An integral transform 114
B_2. Partition of unity 115
C. Bishop's localization theorem and applications 116
C_1. The localization theorem 116
C_2. Applications of Bishop's theorem 118
D. Vitushkin's theorem; a report 121
Remarks about §3 122

§4. Roth's fusion lemma 122
A. The fusion lemma 123
B. A new proof of Bishop's theorem 127
Remark about §4 129

Chapter IV. Approximation on closed sets 130

§1. Uniform approximation by meromorphic functions 130
A. Statement of the problem 130
B. Roth's approximation theorem 131
C. Special cases of the approximation theorem 133
C_1. The one-point compactification G^* of G; connectedness of $G^*\setminus F$ 133
C_2. Three sufficient criteria for meromorphic approximation 134
D. Characterization of the sets, where meromorphic approximation is possible 136

§2. Uniform approximation by analytic functions 136
A. Moving the poles of meromorphic functions 137
B. Preliminary topological remarks 138
C. Arakeljan's approximation theorem 139
C_1. Approximation of meromorphic functions by analytic functions 140
C_2. Arakeljan's theorem 142
Remarks about §2 144

§3. Approximation with given error functions 145
A. The problem; Carleman's theorem 145
A_1. Tangential approximation; ϵ-approximation 145
A_2. Two lemmas 146
A_3. Carleman's theorem 149

B. The special case where F is nowhere dense 151
B_1. Sufficient conditions for ϵ-approximation 151
B_2. Tangential approximation if $F^\circ = \phi$ 155
C. Nersesjan's theorem 155
C_1. Condition (A); a lemma 155
C_2. Nersesjan's theorem 157
Remarks about §3 159

§4. Approximation with certain error functions 160
A. ϵ-approximation without condition (A) 161
B. Growth of the approximating function 162
C. The special case $F = \mathbb{R}$ 163

§5. Some applications of the approximation theorems 164
A. Radial boundary values of entire functions 164
B. Boundary behavior of functions analytic in the unit desk 168
B_1. A general approximation theorem 168
B_2. The Dirichlet problem for radial limits 170
C. Approximation and uniqueness theorems 171
D. Various further constructions 173
D_1. Prescribed boundary behavior along countably many curves 173
D_2. Analytic functions with prescribed cluster sets ... 174
D_3. Schneider's noodles 175
D_4. Julia directions of entire functions 175
Remarks about §5 176

References 178
Index 193

PREFACE TO THE ENGLISH EDITION

Just about one hundred years have passed since Carl Runge published the first general theorem on complex approximation. The field has developed considerably since then and has attracted many function theorists and mathematicians with applied interests. One of the chief promoters of the field was Professor J.L. Walsh, and I was fortunate to have been able to study and work with him in 1955 as Research Fellow.

In my book I have tried to give a synthesis of the concrete, constructive aspects of complex approximation (expansions, interpolation) with the more theoretical results that are connected with the names Mergelyan, Arakeljan, Roth, and others. The German edition appeared in 1980, and soon afterwards Professors W.H.J. Fuchs, L.N. Trefethen, and L.L. Schumaker proposed that an English edition be prepared. Through the recommendation of Professor Piranian, I was fortunate to find Professor Renate McLaughlin to translate the book and to do the proofreading. I sincerely thank her for the immense amount of work that she has put into the project.

The English translation follows the German original rather closely; however, there are several improvements and additions to the text and about 25 new references. The translation was finished in March 1983, but for various reasons the publication of the translated book was delayed until now. This explains why some important work that has appeared after 1983 is not mentioned here.

I am very grateful that Birkhäuser has agreed to publish an English edition, and I hope that it will attract new friends to the growing field of complex approximation.

Giessen, April 1987 — Dieter Gaier

PREFACE TO THE GERMAN EDITION

The present book essentially consists of two parts that arose at different occasions and that address different interests.

The first part, consisting of Chapters I and II, contains the classical portions of complex approximation. Here the more constructive aspects are emphasized: The approximation of a function by series expansions (in orthogonal polynomials or in Faber polynomials) as well as by interpolation. The basis for this was a one-semester course that I have taught in Giessen several times.

The second part, consisting of Chapters III and IV, originated with lectures that I gave at a tutorial conference in Oberwolfach with the topic, "Complex approximation," and on short visits to Stockholm and Pasadena. The lectures were to give an overview of important developments since Mergelyan's theorem. We first present general theorems about approximation on compact sets by polynomials and rational functions. Then we deal with results about approximation by meromorphic, rational, and analytic functions on compact or only closed sets (in $\mathbb{C}$ or in a general domain G); these are connected with the names of Alice Roth and Arakeljan. The latter results are important for the construction of analytic functions with a prescribed boundary behavior; this topic is dealt with in some detail at the end of the book.

The second part is largely independent of the first, so that the reader interested only in newer developments could begin with Chapter III. However, one should realize that the goal of the book has not been to present all newer results; rather, I have attempted throughout to lead the reader to the newer literature. A detailed bibliography can be found at the end of the book.

Giessen, West Germany
July 1980

Dieter Gaier

SYMBOLS AND NOTATION

$:=$ the symbol on the side of the colon is being defined
$\mathbb{C}$: the set of complex numbers
$\hat{\mathbb{C}} = \mathbb{C} \cup \{\infty\}$: extension of the complex numbers
G : a domain in $\mathbb{C}$
$G^* = G \cup \{\infty\}$: one-point compactification of G
$\overline{G}$: the closure of G in $\mathbb{C}$
M' : the set of limit points of $M \subset \mathbb{C}$
∂M : the set of boundary points of M
$\mathbb{D} = \{z \in \mathbb{C} : |z| < 1\}$
K : a compact set in $\mathbb{C}$
K° : the set of interior points of K
$K^c = \mathbb{C} \setminus K$
F : a closed set (closed in $\mathbb{C}$ or in G)
F° : the set of interior points of F
$\operatorname{dist}(A, B) = \inf\{|z_1 - z_2| : z_1 \in A, z_2 \in B\}$
ϕ : the empty set

Classes of functions:

$C(K)$: the set of functions $f: K \to \mathbb{C}$ continuous on K
$A(K) = \{f \in C(K) : f \text{ analytic in } K^\circ\}$
$R(K) = \{f \in A(K)$: for every $\epsilon > 0$ there exists a rational function R such that $\|f - R\| < \epsilon\}$
$M(G)$: the set of functions meromorphic in G
$\operatorname{Hol}(G)$: the set of functions analytic in G

PART I

APPROXIMATION BY SERIES EXPANSIONS AND BY INTERPOLATION

We begin with the more constructive aspects of approximation theory, namely series expansions of various types and interpolation methods. Later, in Part II, we shall for the most part deal with existence theorems.

Chapter I

REPRESENTATION OF COMPLEX FUNCTIONS BY ORTHOGONAL SERIES AND FABER SERIES

As is well known, one of the most important methods of representing functions defined on real or complex domains with the help of simpler functions is the method of series expansions. The theory of convergence for functions defined on complex domains, especially for analytic functions, is considerably simpler than for functions defined on real domains. Since we are generally interested in analytic functions, we shall mainly be concerned with series developments in the space $L^2(G)$. The first four sections of this chapter are devoted to this topic. An important element in the space $L^2(G)$ is the Bergman kernel function, which is useful for the construction of conformal mappings. We talk about the Bergman kernel function in Section 5. Finally, in Section 6, we present the expansion of functions in Faber polynomials in order to obtain certain theorems on the quality of approximation by polynomials.

References for this chapter are Behnke and Sommer [1962, Chapter III, §12 and 13], Bergman [1970], Epstein [1965], Gaier [1964, Chapter III], Nehari [1952, Chapter V, §10].

§1. The Hilbert space $L^2(G)$

In this section, all series expansions will take place in $L^2(G)$. We introduce this space first and then establish some of its properties.

A. Definition of $L^2(G)$

Suppose $G \subset \mathbb{C}$ is an arbitrary domain, f is analytic in G, and set

$$I[f]: = \iint_G |f(z)|^2 \, dm,$$

where dm is the two-dimensional Lebesgue measure. The integral can also be interpreted as a limit of Riemann integrals, as follows. Suppose $\{G_n\}$ is a sequence of subsets that exhaust G (see, for example, Walsh [1969, p. 7]); this means the sets G_n have the properties that (i) each G_n is a domain whose boundary ∂G_n consists of finitely many Jordan curves; (ii) $G_n \subset G_{n+1} \subset G$ for each n; (iii) for each point $P \in G$ there exists an $n_0 = n_0(P)$ such that $P \in G_n$ for $n > n_0$. If we let

$$\phi_n(z) = \begin{cases} |f(z)|^2 & \text{for } z \in \overline{G}_n, \\ 0 & \text{for } z \in G \backslash \overline{G}_n, \end{cases}$$

we see that $\phi_n \uparrow |f|^2$ in G. The Lebesgue Monotone Convergence Theorem now implies that

$$\iint_G \phi_n \, dm \to \iint_G |f|^2 \, dm \quad (n \to \infty);$$

that is,

$$\iint_{\overline{G}_n} |f|^2 \, dm \to I[f] = \iint_G |f|^2 \, dm \quad (n \to \infty).$$

This establishes $I[f]$ as a limit of Riemann integrals.

We now compute $I[f]$ for a *special case*. Let $G = \{z: r < |z| < R\}$ $(0 \leqslant r < R < \infty)$, and let

$$f(z) = \sum_{n=-\infty}^{\infty} a_n z^n \quad (z \in G).$$

It follows that

$$\begin{aligned} I[f] &= \int_{\rho=r}^{R} \int_{\phi=0}^{2\pi} (\Sigma\, a_n \rho^n e^{in\phi})(\Sigma\, \overline{a}_m \rho^m e^{-im\phi}) \rho \, d\phi \, d\rho \\ &\underset{(a)}{=} 2\pi \int_{\rho=r}^{R} \Sigma\, |a_n|^2 \rho^{2n+1} \, d\rho \\ &\underset{(b)}{=} 2\pi\, \Sigma\, |a_n|^2 \int_r^R \rho^{2n+1} \, d\rho. \end{aligned}$$

Equality occurs at (a) because both series converge absolutely for $r < \rho < R$ and uniformly in ϕ; equality occurs at (b) because the terms of the series are nonnegative.

Corollary for $r = 0$: If f is analytic in the punctured disk $0 < |z| < R$ and if $I[f] < \infty$, then $a_n = 0$ for $n < 0$. In this case, the point $z = 0$ is a removable singularity of f, and we have the relation

(1.1) $$I[f] = \pi \, \Sigma_{n=0}^{\infty} \frac{|a_n|^2}{n+1} R^{2n+2};$$

in other words, $I[f]$ can be represented explicitly in terms of the coefficients of f.

Now let $G \subset \mathbb{C}$ again denote an arbitrary domain.

Definition 1. *Let*

$$L^2(G) = \{f: f \text{ analytic in } G \text{ and } I[f] < \infty\}.$$

This definition is analogous to the corresponding definition for functions defined on real domains; however, here $|f(z)|$ $(z \in G)$ can be estimated by $I[f]$.

Lemma 1. *Suppose* $f \in L^2(G)$, $z \in G$, *and* $d_z = \text{dist}(z, \partial G)$. *Then*

$$|f(z)|^2 \leqslant \frac{I[f]}{\pi d_z^2}. \tag{1.2}$$

Proof. We have that $I[f] \geqslant \iint_D |f|^2\, dm$, where D is the disk with radius d_z and center z. Relation (1.1) implies that

$$\iint_D |f|^2\, dm \geqslant \pi |a_0|^2 R^2 = \pi |f(z)|^2 d_z^2,$$

and inequality (1.2) is established.

In the following, inequality (1.2) will be used repeatedly. It is sharp: equality holds in the case where G is the unit disk, $f = 1$, and $z = 0$. We also note that if $G = \mathbb{C}$, then $L^2(G)$ contains only the function $f = 0$; hence we can exclude the case $G = \mathbb{C}$.

B. $L^2(G)$ as a Hilbert space

The inequality $|a + b|^2 \leqslant 2(|a|^2 + |b|^2)$ implies that

$$\text{(i)} \qquad |af(z) + bg(z)|^2 \leqslant 2(|a|^2 |f(z)|^2 + |b|^2 |g(z)|^2)$$

for any two functions $f, g \in L^2(G)$; further, the identity

$$\text{(ii)} \qquad f\bar{g} = \tfrac{1}{2}|f + g|^2 + \frac{i}{2}|f + ig|^2 - \frac{1+i}{2}|f|^2 - \frac{1+i}{2}|g|^2$$

holds.

Definition 2. *For* $f, g \in L^2(G)$, *we write*

$$(f, g) = \iint_G f(z)\overline{g(z)}\, dm. \tag{1.3}$$

According to relations (i) and (ii), expression (1.3) is a complex number, called the *inner product* of f and g. We now prove the following theorem.

Theorem 1. *With (f, g) defined as in (1.3), $L^2(G)$ is a Hilbert space.*

Proof. We show that $L^2(G)$ is a complete inner-product space. (a) By (i), the space is closed under addition and scalar multiplication. (b) The inner product defined by (1.3) has all required properties, namely,

$$(f+g, h) = (f, h) + (g, h); (f, g) = \overline{(g, f)};$$
$$(af, g) = a(f, g) \quad \text{for } a \in \mathbb{C};$$
$$(f, f) \geqslant 0; \text{ and } (f, f) = 0 \quad \text{if and only if} \quad f = 0.$$

As usual, $L^2(G)$ becomes a normed space if we define

$$\| f \| := (f, f)^{1/2} = \sqrt{\iint_G |f(z)|^2 \, dm}.$$

(c) It remains to be shown that $L^2(G)$ is complete with this norm. Suppose $\{f_n\}$ is a Cauchy sequence in $L^2(G)$; that is,

$$\|f_n - f_m\|^2 = I[f_n - f_m] < \epsilon \quad \text{if } n, m > N.$$

For each compact subset B of G, inequality (1.2) implies that

$$|f_n(z) - f_m(z)|^2 < \frac{\epsilon}{\pi d^2} \qquad (z \in B),$$

where $d = dist(B, \partial G)$. This means that on each compact subset B of G, the sequence $\{f_n\}$ converges uniformly to an analytic function F:

$$f_n(z) \rightrightarrows F(z) \qquad (n \to \infty; z \in B \subset G).$$

The inequality $I[f_n - f_m] < \epsilon$ further implies $\iint_B |f_n - f_m|^2 \, dm < \epsilon$ $(n, m > N)$. If we now let $m \to \infty$, we obtain that $\iint_B |f_n - F|^2 \, dm \leqslant \epsilon$ $(n > N)$ for each compact $B \subset G$; hence $I[f_n - F] \leqslant \epsilon$ $(n > N)$. The last inequality implies that $F \in L^2(G)$ and that $\|f_n - F\| \to 0$ $(n \to \infty)$. In other words, each Cauchy sequence in $L^2(G)$ converges.

We note that the theory of expansions in the space $L^2(G)$ began to be developed about the year 1922 by Bergman, Bochner, and Carleman. Instead of definition (1.3), one can introduce more generally the inner product $(f, g) = \iint_G f\bar{g}w \, dm$ with weight function w, or one can introduce analogous line integrals on ∂G.

§2. Orthonormal systems of polynomials in $L^2(G)$

If H is a Hilbert space (or only a vector space with an inner product), we say that a subset $S \subset H$ is an *ON system* (orthonormal system) in H if

$$(u, v) = \begin{cases} 1 & \text{if } u = v, \\ 0 & \text{if } u \neq v, \end{cases} \qquad \text{whenever } u, v \in S.$$

An important tool for the approximation of elements in H is the method of series expansions with respect to an ON system (see §4). First, we study ON systems themselves.

A. Construction of ON systems; Gramian matrix

Each finite, nonempty subset $\{v_1, \ldots, v_n\}$ of an ON system is linearly independent: An equation

$$c_1 v_1 + \ldots + c_n v_n = 0$$

would imply that the inner product $(c_1 v_1 + \ldots + c_n v_n, v_k)$ is also zero; hence $c_k \cdot 1 = 0$ $(k = 1, 2, \ldots, n)$. Conversely, any linearly independent set with n elements $u_1, \ldots, u_n \in H$ generates an ON system with n elements. Section A_1 gives a recursive method for generating such an ON system, and Section A_2 gives an explicit construction.

A_1. The Gram-Schmidt orthogonalization process

Suppose $\{u_1, \ldots, u_n\} \subset H$ is a linearly independent set with n elements. We construct recursively an ON system $v_1, \ldots, v_n$.

1st step. Let

$$v_1^* = u_1, \quad D_1 = (v_1^*, v_1^*)^{1/2}, \quad v_1 = v_1^*/D_1.$$

kth step $(k = 2, \ldots, n)$. Let

$$v_k^* = u_k - \sum_{j<k} (u_k, v_j) v_j, \quad D_k = (v_k^*, v_k^*)^{1/2}, \quad v_k = v_k^*/D_k.$$

Note that $D_k > 0$ for each k; for if D_k were zero for some k, it would follow that $v_k^* = 0$, and u_k would be a linear combination of $v_1, \ldots, v_{k-1}$ and hence of $u_1, \ldots, u_{k-1}$. This contradicts our assumption about the linear independence of the u_j.

The elements ν_k obviously are normalized: $(\nu_k, \nu_k) = 1$. Further, it is easy to show by induction that ν_k and ν_j $(j < k)$ are orthogonal. Hence the elements $\nu_1, \ldots, \nu_n$ form an ON system.

Note that

$$
\begin{aligned}
&\nu_1 = a_{11}u_1, \\
&\nu_2 = a_{21}u_1 + a_{22}u_2, \\
(2.1)\quad &\ldots \\
&\nu_n = a_{n1}u_1 + a_{n2}u_2 + \ldots + a_{nn}u_n, \\
&\text{where } a_{kk} > 0,
\end{aligned}
\qquad \text{and} \qquad
\begin{aligned}
&u_1 = b_{11}\nu_1, \\
&u_2 = b_{21}\nu_1 + b_{22}\nu_2, \\
&\ldots \\
&u_n = b_{n1}\nu_1 + b_{n2}\nu_2 + \ldots + b_{nn}\nu_n, \\
&\text{where } a_{kk}b_{kk} = 1 \quad (k = 1, \ldots, n).
\end{aligned}
$$

By the way, if one requires that $a_{kk} > 0$ $(k = 1, \ldots, n)$, the linear combinations $\nu_1, \ldots, \nu_n$ in (2.1) are *uniquely determined* by the system $u_1, \ldots, u_n$. For if $\{\nu_k\}$ and $\{\nu'_k\}$ are two such ON systems, the element $\nu_k/a_{kk} - \nu'_k/a'_{kk}$ is a linear combination of $u_1, \ldots, u_{k-1}$ as well as orthogonal to $u_1, \ldots, u_{k-1}$; hence

$$\left(\frac{\nu_k}{a_{kk}} - \frac{\nu'_k}{a'_{kk}}, \frac{\nu_k}{a_{kk}} - \frac{\nu'_k}{a'_{kk}}\right) = 0 \qquad (k = 1, \ldots, n).$$

This, in turn, implies that $\nu_k = C_k\nu'_k$, and $|C_k| = 1$ because ν_k and ν'_k are normalized. The requirements that $a_{kk} > 0$ and $a'_{kk} > 0$ now assure that $C_k = 1$, and therefore $\nu_k = \nu'_k$.

A_2. The Gramian matrix

Definition 1. *Suppose H is a linear space with an inner product and $x_1, \ldots, x_n \in H$ are arbitrary elements. The matrix*

$$G = G(x_1, \ldots, x_n) = \begin{pmatrix} (x_1, x_1) \ldots (x_1, x_n) \\ (x_2, x_1) \ldots (x_2, x_n) \\ \ldots \\ (x_n, x_1) \ldots (x_n, x_n) \end{pmatrix}$$

is called the ***Gramian matrix*** *of $x_1, \ldots, x_n$, and*

$$g = g(x_1, \ldots, x_n) = \det G(x_1, \ldots, x_n)$$

is called the corresponding ***Gramian determinant***.

Remarks. 1. Obviously, G is Hermitian: $\overline{G}' = G$. If $a_j, b_j \in \mathbb{C}$ and we define the vectors $a = (a_1, \ldots, a_n)'$ and $b = (b_1, \ldots, b_n)'$, then

$$\left(\sum_{i=1}^{n} a_i x_i, \sum_{j=1}^{n} b_j x_j\right) = \sum_{i,j=1}^{n} a_i \bar{b}_j (x_i, x_j) = a' G \bar{b}.$$

If we choose $b = a$, we see that G is positive semi-definite. It follows that the Gramian determinant g is nonnegative, and $g = 0$ if and only if the quadratic form $a'G\bar{a}$ vanishes for some $a \neq 0$. Hence $g = 0$ if and only if the set $\{x_1, \ldots, x_n\}$ is linearly dependent.

2. If $H = \mathbf{R}^n$, then g has a geometric interpretation. Let $x_j = (x_{j1}, x_{j2}, \ldots, x_{jn}) \in \mathbf{R}^n$ $(j = 1, \ldots, n)$, and form the coordinate matrix

$$M = \begin{pmatrix} x_{11} & x_{12} & \cdots & x_{1n} \\ \cdots & & \cdots & \cdots \\ x_{n1} & x_{n2} & \cdots & x_{nn} \end{pmatrix}.$$

Then $G = MM'$, and hence $g = \det G = (\det M)^2 = V^2$, where V is the volume of the parallelepiped spanned by $x_1, \ldots, x_n$ in $\mathbf{R}^n$.

Now suppose we again have a linearly independent set of n elements $u_1, \ldots, u_n \in H$. We wish to construct an ON system $v_1, \ldots, v_n$. We set

$$A_1 = (u_1, u_1), \quad v_1 = u_1/(A_1)^{1/2},$$

and for $k = 2, \ldots, n$, we set

$$(2.2) \qquad A_k = g(u_1, \ldots, u_k) = \det \begin{pmatrix} (u_1, u_1) \ldots (u_1, u_k) \\ \ldots\ldots\ldots\ldots\ldots\ldots \\ (u_k, u_1) \ldots (u_k, u_k) \end{pmatrix} > 0$$

and

$$(2.3) \qquad v_k^* = \det \begin{pmatrix} (u_1, u_1) \ldots (u_1, u_{k-1}) & u_1 \\ \ldots\ldots\ldots\ldots\ldots\ldots & \\ (u_k, u_1) \ldots (u_k, u_{k-1}) & u_k \end{pmatrix}.$$

Clearly, v_k^* is a linear combination of $u_1, \ldots, u_k$. It follows immediately that $(v_k^*, u_j) = 0$ $(j < k)$ and hence $(v_k^*, v_j^*) = 0$ $(j < k)$. Further,

$$(v_k^*, v_k^*) = \det \begin{pmatrix} (u_1, u_1) \ldots (u_1, u_{k-1}) & (u_1, v_k^*) \\ \ldots\ldots\ldots\ldots\ldots\ldots & \\ (u_k, u_1) \ldots (u_k, u_{k-1}) & (u_k, v_k^*) \end{pmatrix}.$$

Since $(u_k, v_k^*) = \overline{(v_k^*, u_k)} = \bar{A}_k = A_k$ and the other entries in the last column are zero, we see that

$$(v_k^*, v_k^*) = A_k \cdot A_{k-1}.$$

Result. If A_k and v_k^* are defined as in (2.2) and (2.3), the elements

$$v_1 = u_1/(A_1)^{1/2}, \quad v_k = v_k^*/(A_k A_{k-1})^{1/2} \quad (k = 2, \ldots, n)$$

form an ON system.

Finally, we note that in the expression

$$v_k = a_{k1}u_1 + \ldots + a_{kk}u_k$$

the coefficient a_{kk} is given by $a_{kk} = \sqrt{A_{k-1}/A_k}$.

A_3. A special case: Polynomials in $L^2(G)$

We return to our special Hilbert space $L^2(G)$ and choose $u_j = z^{j-1}$ ($j = 1, 2, \ldots$). If G is bounded, the u_j clearly belong to $L^2(G)$, and each finite subset of $\{u_j\}$ is linearly independent. We can therefore use one of the procedures mentioned in Sections A_1 and A_2 to construct an orthonormal system. As the $(n + 1)$st element of the ON system we obtain a uniquely determined polynomial of degree n, namely,

$$P_n(z) = C_0^{(n)} + C_1^{(n)}z + \ldots + C_n^{(n)}z^n,$$

where

$$k_n := C_n^{(n)} > 0 \quad (n = 0, 1, 2, \ldots).$$

For the practical determination of P_n, both methods above require the evaluation of the double integrals (z^p, z^q), where $p, q = 0, 1, 2, \ldots$. These double integrals can be converted into one-dimensional integrals as outlined in Sections a) and b) below.

a) G is starlike with respect to 0. If the boundary of G is a Jordan curve and the equation $r = r(\phi)$ $(0 \leqslant \phi \leqslant 2\pi)$ represents ∂G in polar coordinates, the integral becomes

$$(z^p, z^q) = \iint_G z^p \overline{z^q}\, dm = \int_{\phi=0}^{2\pi} \int_{\rho=0}^{r(\phi)} \rho^{p+q+1} e^{i\phi(p-q)}\, d\rho\, d\phi$$

$$= \frac{1}{p+q+2} \int_{\phi=0}^{2\pi} [r(\phi)]^{p+q+2} e^{i\phi(p-q)}\, d\phi.$$

For large values of p-q, the integrand strongly oscillates with ϕ and can create numerical difficulties; it may be necessary to use double precision.

b) Use of Green's formula. Suppose G is a simply or multiply connected domain with piecewise smooth, positively oriented boundary ∂G, and suppose the functions f and g are analytic in G with continuous derivatives in $\overline{G}$. An application of Green's formula leads to

$$\iint_G f\overline{g'}\,dm = \frac{1}{2i}\int_{\partial G} f\overline{g}\,dz.$$

Applying this to the inner products (z^p, z^q), one obtains

$$\iint_G z^p\overline{z^q}\,dm = \frac{1}{2i(q+1)}\int_{\partial G} z^p\overline{z^{q+1}}\,dz.$$

Again, only a one-dimensional integral needs to be evaluated. If ∂G is a polygon, then on each of its sides $dz = c\,ds$, and Gaussian quadrature can be used to evaluate the integral.

We conclude with a simple *example.* Let $G = \{z\colon |z| < 1\}$. Both formulas above give

$$(z^p, z^q) = \begin{cases} 0 & \text{if } p \neq q, \\ \dfrac{\pi}{q+1} & \text{if } p = q. \end{cases}$$

Hence

$$A_k = \det\begin{pmatrix} \pi & 0 & \dots & 0 \\ 0 & \pi/2 & \dots & 0 \\ \dots & \dots & \dots & \dots \\ 0 & 0 & \dots & \pi/k \end{pmatrix} = \pi^k/k!$$

and

$$v_k^* = \det\begin{pmatrix} \pi & 0 & \dots & 0 & 1 \\ 0 & \pi/2 & \dots & 0 & z \\ \dots & \dots & \dots & \dots & \dots \\ 0 & 0 & \dots & \pi/(k-1) & z^{k-2} \\ 0 & 0 & \dots & 0 & z^{k-1} \end{pmatrix} = \frac{\pi^{k-1}}{(k-1)!}\,z^{k-1}.$$

Therefore the in $L^2(G)$ orthonormal polynomials are $v_k = \sqrt{k/\pi}\,z^{k-1}$ $(k = 1, 2, \dots)$.

B. Zeros of orthogonal polynomials

Real polynomials that are orthogonal with respect to an inner product $(f, g) = \int_a^b fgw\,dx$ have important properties (see, for example, Davis [1963, p. 234 ff.]): They satisfy a three-term recursion formula, all zeros are simple, and the zeros lie in the interval (a, b). For our ON polynomials in $L^2(G)$, unfortunately, the first property is valid only in exceptional cases, and the last example in Section A_3 exhibits ON polynomials with multiple zeros at the origin. However, we do have the following beautiful theorem about the location of the zeros.

Theorem 1 (Fejér 1922). *All zeros of the ON polynomials P_n lie in the convex hull of G.*

The proof uses the fact that the P_n are closely related to certain minimal polynomials p_n. Let

$$K_n = \{q: q(z) = c_0 + c_1 z + \ldots + c_{n-1} z^{n-1} + z^n\}.$$

Lemma 1. *The polynomial q has minimal norm in K_n if and only if $q = p_n := P_n/k_n$, where k_n is the leading coefficient of P_n.*

Proof. Since $q - P_n/k_n$ has degree $n - 1$ and can therefore be expressed as a linear combination of the polynomials P_j $(j < n)$, the equations

$$\begin{aligned}(q - P_n/k_n, q - P_n/k_n) &= \|q\|^2 - (P_n, q)/k_n + 0 \\ &= \|q\|^2 - (P_n, z^n)/k_n\end{aligned}$$

hold for each $q \in K_n$. The assertion follows.

To establish Theorem 1, it now suffices to show that all zeros of the polynomials p_n lie in the convex hull G^* of G. Write

$$p_n(z) = (z - z_1) \ldots (z - z_n),$$

and assume that, say, $z_1 \notin G^*$. Thus there exists a line L separating z_1 from G. Let z_1' denote the intersection of L and the perpendicular through z_1. Obviously,

$$|z - z_1'| < |z - z_1| \quad \text{for all } z \in G.$$

The polynomial $q(z) = (z - z_1')(z - z_2) \ldots (z - z_n) \in K_n$ now satisfies the inequality $\|q\| < \|p_n\|$, which contradicts the minimality of p_n. This completes the proof of Theorem 1.

C. Asymptotic representation of the ON polynomials

We shall now study the asymptotic behavior (as $n \to \infty$) of the ON polynomials P_n and of their leading coefficients k_n for the case where ∂G is an analytic Jordan curve. This is important, because it allows us to approximate the conformal mapping of the exterior of ∂G and the capacity of ∂G.

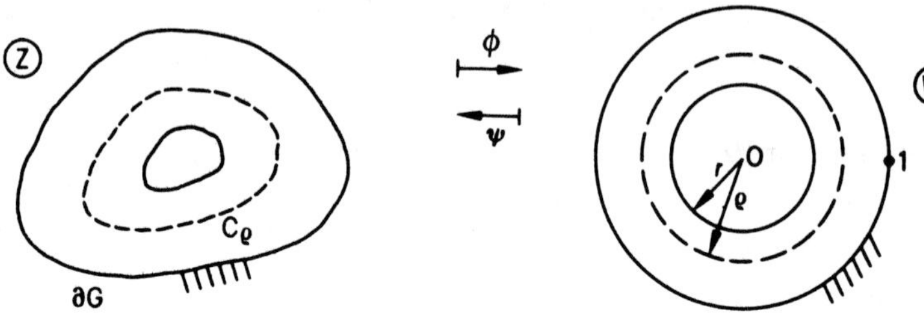

Suppose $z = \psi(w) = cw + c_0 + \frac{c_1}{w} + \ldots$ is the conformal mapping of $\{w\colon |w| > 1\}$ onto the exterior of ∂G, normalized at ∞ such that $c > 0$. The quantity c is called the *capacity* of ∂G. Since ∂G is assumed to be an analytic Jordan curve, there exists an analytic and univalent continuation of ψ to the domain $\{w\colon |w| > r\}$ for some $r < 1$. Let ϕ denote the inverse mapping, and set $C_\rho = \{z\colon |\phi(z)| = \rho, \rho > r\}$.

Theorem 2 (Carleman 1922). *With the notation above, we have*

$$k_n = \sqrt{\frac{n+1}{\pi}}\, c^{-n-1}[1 + O(r^{2n})] \qquad (n \to \infty) \tag{2.4}$$

and

$$P_n(z) = \sqrt{\frac{n+1}{\pi}}\, \phi'(z)[\phi(z)]^n [1 + A_n], \tag{2.5}$$

where

$$A_n = \begin{cases} O(\sqrt{n})r^n & \text{if } z \in C_\rho, \rho \geqslant 1, \\ O(1/\sqrt{n})(r/\rho)^n & \text{if } z \in C_\rho, r < \rho < 1. \end{cases}$$

Even if ∂G is not analytic, it is possible to make statements of this nature (Suetin [1969], [1971], [1972]), but they are more difficult to prove. Before we prove Theorem 2, we draw some *conclusions*.

1. Relation (2.4) implies that

$$\sqrt{\frac{n+1}{n+2}} \cdot \frac{k_{n+1}}{k_n} = c^{-1} + O(r^{2n}) \qquad (n \to \infty).$$

In other words, the capacity c of ∂G can be computed from the leading coefficients k_n of P_n.

2. Relation (2.5) implies that

$$\sqrt{\frac{n+1}{n+2}} \cdot \frac{P_{n+1}(z)}{P_n(z)} = \phi(z) + \begin{cases} O(\sqrt{n})r^n & z \in C_\rho, \rho \geqslant 1, \\ O(1/\sqrt{n})(r/\rho)^n & z \in C_\rho, r < \rho < 1. \end{cases}$$

In other words, the conformal mapping ϕ can be computed from the ON polynomials P_n.

3. In G, the polynomials P_n tend to zero rapidly:

$$\max \{|P_n(z)|: z \in C_\rho, r < \rho < 1\} = O(\sqrt{n} \cdot \rho^n) \qquad (n \to \infty).$$

This estimate cannot be improved, as the ON polynomials in the unit disk demonstrate.

4. Relation (2.5) further shows that the exterior of each C_ρ $(\rho > r)$ contains at most finitely many zeros of the polynomials P_n.

Proof of Theorem 2. We shall use the minimal polynomials p_n with norm $\|p_n\| = (k_n)^{-1}$ that were introduced in Section B. Note that $P_n = k_n p_n$.

1st step. Suppose q is an arbitrary element in $K_n = \{q: q(z) = c_0 + c_1 z + \ldots + c_{n-1}z^{n-1} + z^n\}$, and let G_ρ denote the ring domain

$$G_\rho = \{z: \rho < |\phi(z)| < 1\} \qquad (r < \rho < 1).$$

Using Green's formula, we evaluate the following two integrals:

$$I[q] = \iint_G |q|^2 dm \quad \text{and} \quad I_\rho[q] = \iint_{G_\rho} |q|^2\, dm \quad (q \in K_n).$$

If r is a polynomial with $r' = q$, Green's formula implies that

$$I[q] = \frac{1}{2i} \int_{\partial G} q\bar{r}\, dz = \frac{1}{2i} \int_{|w|=1} F'(w)\overline{F(w)}\, dw,$$

where

$$F(w) = r(\psi(w)) = c^{n+1}\Big(\frac{w^{n+1}}{n+1} + A_0 + A_1 w + \ldots + A_n w^n + \Sigma_{j=1}^{\infty} a_j w^{-j}\Big)$$

$$(|w| > r).$$

The coefficients a_j and A_j depend on the coefficients of q. We observe that the n coefficients $A_1, \ldots, A_n$ can be made to vanish by an appropriate choice of the n coefficients of q; suppose $q = q_0$ represents such a choice. Integration over the unit circle $|w| = 1$ now yields

$$I[q] = \pi c^{2n+2}\left(\frac{1}{n+1} + \Sigma_1^n j|A_j|^2 - \Sigma_1^\infty j|a_j|^2\right). \tag{2.6}$$

In a similar way we obtain

$$I_\rho[q] = \pi c^{2n+2}\left[\frac{1-\rho^{2n+2}}{n+1} + \Sigma_1^n j|A_j|^2(1-\rho^{2j}) + \Sigma_1^\infty j|a_j|^2(\rho^{-2j}-1)\right] \tag{2.7}$$

for each $q \in K_n$.

2nd step. We now prove relation (2.4). On the one hand, equation (2.6) implies

$$(k_n)^{-2} = I[p_n] \leqslant I[q_0] \leqslant \frac{\pi}{n+1}c^{2n+2} \qquad (n = 1, 2, \ldots);$$

on the other hand, equation (2.7) leads to

$$(k_n)^{-2} = I[p_n] \geqslant I_\rho[p_n] \geqslant \frac{\pi}{n+1}c^{2n+2}(1-\rho^{2n+2}) \qquad (n = 1, 2, \ldots).$$

Using both estimates together and letting $\rho \to r$, we obtain

$$\sqrt{\frac{n+1}{\pi}}\,c^{-n-1} \leqslant k_n \leqslant \sqrt{\frac{n+1}{\pi}}\,c^{-n-1}(1-r^{2n+2})^{-1/2} \qquad (n = 1, 2, \ldots),$$

and (2.4) follows.

3rd step. In this step, we prove an inequality for the coefficients a_j and A_j corresponding to the choice $q = p_n$. First note that

$$I_\rho[p_n] \leqslant I[p_n] \leqslant \frac{\pi}{n+1}c^{2n+2}.$$

This, together with (2.7), yields that

$$\Sigma_1^n j|A_j|^2(1-\rho^{2j}) + \Sigma_1^\infty j|a_j|^2(\rho^{-2j}-1) \leqslant \frac{\rho^{2n+2}}{n+1}.$$

If we now let $\rho \to r$ and use the inequalities

$$1 - r^{2j} \geqslant 1 - r^2,\; r^{-2j} - 1 \geqslant r^{-2j}(1-r^2) \qquad (j = 1, 2, \ldots),$$

we obtain

$$(2.8)\qquad \Sigma_1^n j|A_j|^2 + \Sigma_1^\infty j|a_j|^2 r^{-2j} \leqslant \frac{r^{2n+2}}{(n+1)(1-r^2)} \qquad (n = 1, 2, \ldots).$$

4th step. Now we can study the asymptotic behavior of the polynomials p_n. Let $z \in C_\rho (\rho > r)$, so that $|w| = |\phi(z)| = \rho$. The function F corresponding to p_n satisfies the condition $F'(w) = p_n(\psi(w)) \cdot \psi'(w)$, so that $p_n(z) = F'(\phi(z)) \cdot \phi'(z)$. Moreover, we know that

$$F'(w) = c^{n+1}(w^n + \Sigma_1^n jA_j w^{j-1} - \Sigma_1^\infty j a_j w^{-j-1}) = c^{n+1} w^n [1 + \omega(z)],$$

where

$$\omega(z) = \Sigma_1^n jA_j w^{j-1-n} - \Sigma_1^\infty j a_j w^{-j-1-n} \qquad (w = \phi(z)).$$

Substituting into the expression for $p_n(z)$, we find that

$$(2.9)\qquad p_n(z) = c^{n+1}\phi'(z)[\phi(z)]^n [1 + \omega(z)].$$

It remains to estimate $|\omega(z)|$ for $z \in C_\rho$, that is, for $|w| = \rho$.

We have

$$|\omega(z)| \leqslant \Sigma_1^n j|A_j|\rho^{j-1-n} + \Sigma_1^\infty j|a_j|\rho^{-j-1-n} = C_n + D_n,$$

where

$$C_n = \Sigma_1^n \sqrt{j}|A_j|\sqrt{j}\,\rho^{j-1-n} \leqslant (\Sigma_1^n j|A_j|^2)^{1/2} \cdot (\Sigma_1^n j\, \rho^{2j-2-2n})^{1/2}.$$

Clearly, the second factor is $O(n)$ for $\rho \geqslant 1$, but $O(\rho^{-n})$ for $r < \rho < 1$. We estimate the first factor using (2.8) and find that

$$C_n = \begin{cases} O(\sqrt{n}\, r^n) & \text{if } \rho \geqslant 1, \\ O[\frac{1}{\sqrt{n}} (r/\rho)^n] & \text{if } r < \rho < 1. \end{cases}$$

Similarly, we obtain

$$D_n = \rho^{-n-1} \Sigma_1^\infty \sqrt{j}|a_j| \sqrt{j}\, \rho^{-j} = \rho^{-n-1} \Sigma_1^\infty \sqrt{j}\, |a_j| r^{-j} \cdot \sqrt{j} (\frac{r}{\rho})^j$$

$$\leqslant \rho^{-n-1} (\Sigma_1^\infty j|a_j|^2 r^{-2j})^{1/2} (\Sigma_1^\infty j(r/\rho)^{2j})^{1/2} \leqslant (r/\rho)^n \cdot \frac{1}{\sqrt{n}} \cdot O(1) \qquad (n \to \infty).$$

Putting this together, we see that

$$|\omega(z)| \leqslant \begin{cases} O(\sqrt{n}\, r^n) & \text{if } z \in C_\rho \text{ and } \rho \geqslant 1, \\ O[\dfrac{1}{\sqrt{n}} (r/\rho)^n] & \text{if } z \in C_\rho \text{ and } r < \rho < 1. \end{cases}$$

Considering $P_n = k_n p_n$, (2.9), and (2.4), we see that assertion (2.5) follows.

Remark about §2

If one uses a weight function to define the norm in $L^2(G)$, different ON polynomials will result. Their asymptotic behavior and that of their leading coefficients can be described similarly to our Theorem 2 (see the survey article by Suetin [1971]).

§3. Completeness of the polynomials in $L^2(G)$

We shall now explore the question of which assumptions on G will assure that polynomials are dense in $L^2(G)$. This question is important for §4, where we shall deal with the representation of an arbitrary function $f \in L^2(G)$ in terms of the ON polynomials P_n.

A. The problem and examples

Suppose S is a subset of a linear space H with an inner product. Recall that S is *complete* if whenever $y \in H$ and $(x,y) = 0$ for all $x \in S$, it follows that $y = 0$. The set S is *closed* if the linear combinations of elements in S are dense in H. If the space H itself is complete (and therefore a Hilbert space), the two concepts are equivalent.

We now return to our Hilbert space $L^2(G)$ and consider in particular $S = \{1, z, z^2, \ldots\}$ (clearly, $S \subset L^2(G)$ if G is bounded). The question is: When are the polynomials dense in $L^2(G)$?

Definition 1. *A domain $G \subset \mathbb{C}$ has the PA property if the polynomials are dense in $L^2(G)$.*

Here PA stands for polynomial approximation; see Shapiro [1967].

We begin with two simple *examples.* 1. Suppose $G = \{z\colon |z| < 1\}$. Then $f(z) = \Sigma_0^\infty a_k z^k$ belongs to $L^2(G)$ if and only if $\|f\|^2 = \pi\, \Sigma_0^\infty \dfrac{|a_k|^2}{k+1} < \infty$. If we write $P_n(z) = \Sigma_{k \leqslant n}\, a_k z^k$, the latter formula implies that

$$\|f - P_n\|^2 = \pi\, \Sigma_{k>n} \frac{|a_k|^2}{k+1} \to 0 \quad (n \to \infty).$$

Hence G has the PA property.

2. *Multiply connected domains* (with nondegenerate boundary components) *do not have the PA property.* For if G is such a multiply connected domain, there exists a Jordan curve $\Gamma \subset G$ that encloses one of the boundary components of G. We choose four distinct points z_1, z_2, z_3, z_4 on this boundary component and consider the function

$$f(z) = [(z - z_1)(z - z_2)]^{-1/2} \cdot [(z - z_3)(z - z_4)]^{-1/2}.$$

Clearly, f is analytic in G, and in addition we have

$$\iint_G |f(z)|^2 \, dm \leqslant \iint_{\mathbb{C}} \frac{dm}{|z - z_1|\,|z - z_2|\,|z - z_3|\,|z - z_4|} < \infty,$$

so that $f \in L^2(G)$. If there were polynomials P_n with $\|f - P_n\| \to 0$ $(n \to \infty)$, Lemma 1 in §1 would imply that $P_n(z) \rightrightarrows f(z)$ $(n \to \infty)$ on each compact subset of G, hence also on Γ. But then f would have to be analytic in the interior of Γ, which obviously is not the case.

The second example shows that we can restrict the discussion of the PA property to simply connected domains. There is no purely geometric characterization of domains with the PA property; but in the following we give sufficient conditions for G to have (respectively, not to have) the PA property.

B. Domains with the PA property

The following theorem contains the most important sufficient condition for G to have the PA property.

Theorem 1 (Farrell [1934], Markushevich [1934]). *If G is a bounded, simply connected domain whose boundary ∂G is also the boundary of an unbounded domain, then G has the PA property.*

All Jordan domains satisfy the hypotheses of Theorem 1, also domains consisting, for example, of a snake wound infinitely often around the outside of a circle and approaching this circle ("outer snake"), but not a snake wound infinitely often inside a circle and approaching it from the inside ("inner snake"). Domains satisfying the hypotheses of Theorem 1 are also called *Carathéodory domains.*

For the proof, the following three auxiliary results are required.

1. For each Carathéodory domain G there exists a sequence $\{G_n\}$ of Jordan domains such that

$$G \subset G_n \quad \text{and} \quad \overline{G_{n+1}} \subset G_n \qquad (n = 1, 2, \ldots)$$

as well as $G_n \to G$ $(n \to \infty)$. The last condition does not mean that $\cap G_n = G$; instead, G is the largest domain contained in each G_n and containing a fixed point $\zeta \in G$. The domain G is the so-called "kernel" of the sequence $\{G_n\}$. The boundaries of the domains G_n could, for example, be level curves in the exterior of G. The example of the outer snake shows that the kernel of $\{G_n\}$ need not be the same as $\cap G_n$.

2. If h_n is the conformal mapping of the domain G_n onto G, normalized by the conditions $h_n(\zeta) = \zeta$, $h_n'(\zeta) > 0$, then

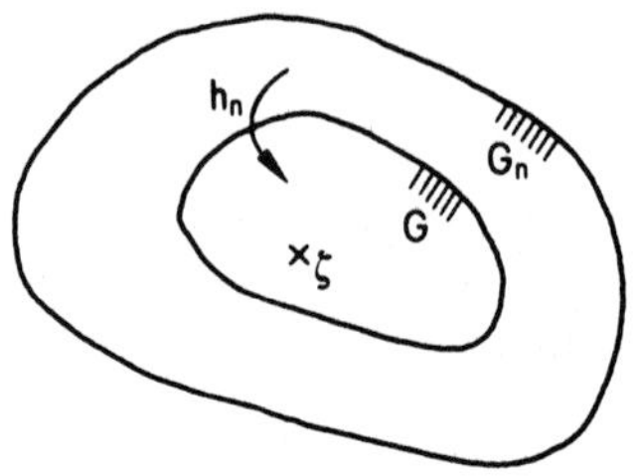

$$h_n(z) \Rightarrow z \quad \text{and} \quad h_n'(z) \Rightarrow 1 \qquad (n \to \infty)$$

uniformly on each compact subset B of G; see, for example, Goluzin [1969, p. 55].

3. Finally, we need a simple version of Runge's theorem, which later will be proved in several different ways. Suppose G is a Jordan domain and f is analytic in $\overline{G}$. Then for each $\epsilon > 0$ there exists a polynomial P such that

$$|f(z) - P(z)| < \epsilon \qquad (z \in \overline{G}).$$

Proof of Theorem 1. Suppose f is an arbitrary function in $L^2(G)$.

1st step. Construction of a function F analytic in $\overline{G}$ such that $\|f - F\| < \epsilon$. Using the functions h_n from above, we define

$$f_n(z) := f(h_n(z)) \cdot h_n'(z) \qquad (z \in G_n).$$

First we show that

$$\iint_G |f_n|^2 \, dm \to \iint_G |f|^2 \, dm \qquad (n \to \infty). \tag{3.1}$$

If we set $w = h_n(z)$, we obtain on the one hand

$$\iint_G |f_n|^2 \, dm = \iint_{h_n(G)} |f|^2 \, dm \leqslant \iint_G |f|^2 \, dm \quad \text{for each } n.$$

On the other hand, $f_n(z) \Rightarrow f(z)$ $(n \to \infty)$ on each compact subset $B \subset G$, and therefore

$$\underline{\lim} \iint_G |f_n|^2 \, dm \geqslant \underline{\lim} \iint_B |f_n|^2 \, dm = \iint_B |f|^2 \, dm.$$

The last statement holds for each compact B; consequently,

$$\iint_G |f|^2\, dm \leqslant \underline{\lim} \iint_G |f_n|^2\, dm \leqslant \overline{\lim} \iint_G |f_n|^2\, dm \leqslant \iint_G |f|^2\, dm.$$

Relation (3.1) is now established, and it implies that

$$\iint_{G\setminus B} |f_n|^2\, dm \to \iint_{G\setminus B} |f|^2\, dm \qquad (n \to \infty)$$

for each compact subset B of G.

Now let $0 < \delta < \epsilon^2/7$ and choose B such that $\iint_{G\setminus B} |f|^2 dm < \delta$. Let n be so large that

$$\iint_{G\setminus B} |f_n|^2\, dm < 2\delta \quad \text{and} \quad \iint_B |f_n - f|^2\, dm < \delta.$$

For this choice of n, we see that

$$\|f_n - f\|^2 = \iint_B |f_n - f|^2\, dm + \iint_{G\setminus B} |f_n - f|^2\, dm$$
$$< \delta + 2\Big(\iint_{G\setminus B} |f_n|^2 dm + \iint_{G\setminus B} |f|^2 dm\Big) < 7\delta < \epsilon^2,$$

and the function $F = f_n$ has the desired properties.

2nd step. Construction of a polynomial P with $\|f - P\| < 2\epsilon$.

We apply Runge's theorem to F and see that for each $\delta > 0$ there exists a polynomial P such that

$$|F(z) - P(z)| < \delta \qquad (z \in \overline{G}),$$

and consequently we have

$$\iint_G |F - P|^2\, dm < \delta^2 \cdot (\text{area of } G) < \epsilon^2$$

for an appropriate choice of δ. In other words, we found a polynomial P such that $\|F - P\| < \epsilon$. The assertion now follows.

The property of being a Carathéodory domain is only sufficient for G to have the PA property. Other more complicated conditions are known; see the survey article by Mergelyan [1953, p. 130] and his articles [1955], [1956]. See also Smirnov and Lebedev [1968, p. 271], Farrell [1966], and Hedberg [1965], [1969]. In these articles the PA property is studied also when a weight function is used.

C. Domains not having the PA property

Typical simply connected domains to which Theorem 1 cannot be applied are the following:

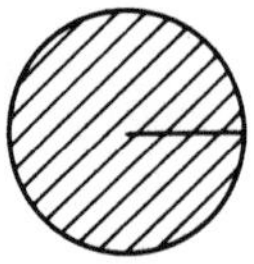

Slit domain

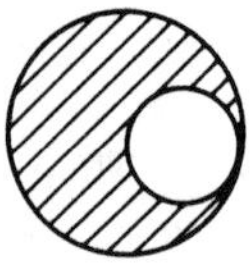

Moon-shaped domain

Inner snake

We deal here with slit domains and moon-shaped domains. Concerning inner snakes, see Remark 5 at the end of §3.

C_1. Slit domains

We use the following elementary result.

Lemma 1. *Suppose G and G' are two domains such that $G' \supset G$ and* meas $(G' \backslash G) = 0$. *If G has the PA property, then G' also has the PA property.*

Proof. Obviously, if $f \in L^2(G')$, then also $f \in L^2(G)$. Since G has the PA property, it follows that for each $\epsilon > 0$ there exists a polynomial P such that $\|f - P\|_G < \epsilon$. Hence $\|f - P\|_{G'} < \epsilon$, because the measure of $G' \backslash G$ is zero.

Consequently, if G' fails to have the PA property, then G will also fail to have it:

Corollary. *G does not have the PA property if there exists a domain G' such that $G' \supset G$,* meas $(G' \backslash G) = 0$, *and G' is multiply connected.*

For example, *no* domain G whose boundary ∂G contains a slit can have the PA property, because a part of the slit can be filled in, and the Corollary can then be applied to the resulting domain G'.

C_2. Moon-shaped domains

As preparation for dealing with general moon-shaped domains, we prove the following lemma.

Lemma 2. *Suppose Γ is a rectifiable Jordan curve and G is the interior of Γ. Suppose F is analytic in G and continuous in $\overline{G}$, and suppose B is a compact subset of G. Then, for each $\alpha > 0$, there exists a constant $M(\alpha, B)$ such that*

$$\max\{|F(z)|: z \in B\} \leqslant M(\alpha, B)\left\{\int_\Gamma |F(z)|^\alpha \, |dz|\right\}^{1/\alpha}. \tag{3.2}$$

Proof. If F has no zeros in G or if $\alpha \geqslant 1$, inequality (3.2) follows immediately from an application of the Cauchy integral formula to F^α or F, respectively. In the general case, we reason as follows.

Suppose ϕ is a conformal mapping of $\{w: |w| < 1\}$ onto G and $\rho < 1$ is chosen such that B lies inside the level curve $\Gamma_\rho = \{z = \phi(w): |w| = \rho\}$. We consider the function

$$F(\phi(w)) \cdot \phi'(w)^{1/\alpha} = B(w) \cdot H(w) \qquad (|w| < 1),$$

where the Blaschke product B absorbs all zeros of $F(\phi(w))$ in the unit disk and the function H does not vanish. For $|w| \leqslant \rho < 1$ we find that

$$\begin{aligned} |F(\phi(w)) \cdot \phi'(w)^{1/\alpha}| &< |H(w)| = |H(w)^\alpha|^{1/\alpha} \\ &\leqslant \Big[\frac{1}{2\pi} \cdot \frac{1}{1-\rho} \int_{|w|=1} |H(w)|^\alpha |dw|\Big]^{1/\alpha}, \end{aligned}$$

and the last integral equals

$$\int_{|w|=1} |H(w)B(w)|^\alpha |dw| = \int_{|w|=1} |F(\phi(w))|^\alpha |\phi'(w)|\, |dw| = \int_\Gamma |F(z)|^\alpha |dz|.$$

If $z \in B$, then $|w| \leqslant \rho$, and thus $|\phi'(w)| \geqslant c(\rho) > 0$. In equality (3.2) now follows.

Now we are ready to discuss moon-shaped domains.

Definition 2. *A (general) moon-shaped domain is a bounded domain G whose boundary consists of two Jordan curves having exactly one point in common.*

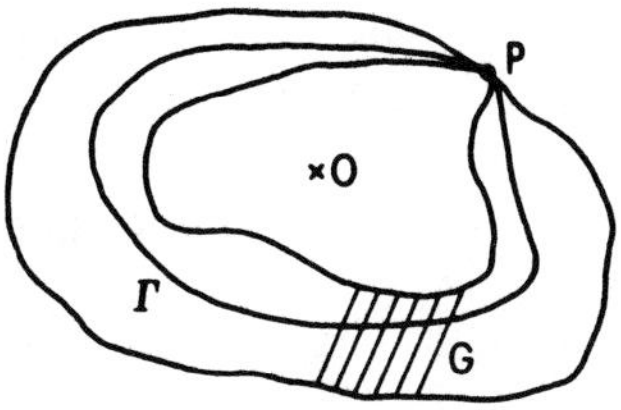

Keldysh [1939] was the first to study the PA property of moon-shaped domains, and he observed that *a domain G has the PA property if and only if the function* $1/\sqrt{z}$ *can be approximated arbitrarily well by polynomials.* Here we have assumed that the origin lies inside the inner Jordan curve of the boundary ∂G.

The stated condition obviously is necessary, because $1/\sqrt{z} \in L^2(G)$. To show that it is also sufficient, suppose f is an arbitrary function in $L^2(G)$. The mapping $w = \sqrt{z}$ transforms G into a Jordan domain G_w and

$$\iint_G |f(z)|^2\, dm = 4 \iint_{G_w} |f(w^2)w|^2\, dm_w < \infty;$$

that is, $wf(w^2) \in L^2(G_w)$. Since polynomials are dense in $L^2(G_w)$, there exists for each $\epsilon > 0$ a polynomial P such that

$$4\iint_{G_w} |wf(w^2) - P(w)|^2 \, dm_w < \epsilon,$$

and therefore $\iint_G |f(z) - P(\sqrt{z})/\sqrt{z}\,|^2 \, dm < \epsilon$. The polynomial P can be separated into its even and odd parts: $P(\sqrt{z}) = P_1(z) + \sqrt{z}\,P_2(z)$. In order to obtain the estimate $\|f - Q\| < 2\epsilon$ for some polynomial Q, it is now sufficient to know that $\inf_R \|1/\sqrt{z} - R\| = 0$ for polynomials R. The last condition then implies that $\inf_R \|P_1/\sqrt{z} - R\| = 0$. Hence Keldysh's condition is also sufficient for G to have the PA property.

Moon-shaped domains, although topologically equivalent to each other, behave rather differently with respect to the PA property. They can have the PA property, as Keldysh has shown by an example ([1939, p. 398]; see also Mergelyan [1953, p. 116]); however, often this is not the case.

Theorem 2. *Suppose there exists a rectifiable Jordan curve* Γ *in* $G \cup \{P\}$ (see the figure following Definition 2) *such that the distance function* $d_z = \operatorname{dist}(z, \partial G)$ $(z \in \Gamma)$ *satisfies the condition*

$$\int_\Gamma \frac{|dz|}{d_z^\alpha} =: I < \infty \tag{3.3}$$

for some $\alpha > 0$. *Then the moon-shaped domain* G *does not have the PA property.*

Remarks. 1. In order for condition (3.3) to be satisfied, the two Jordan curves cannot approach each other too rapidly at the point P. If G is bounded by two circles touching at $z = 1$, then $d_z \approx |z - 1|^2$, and (3.3) is satisfied for each $\alpha < ½$. The moon-shaped domain depicted at the beginning of Section C therefore does not have the PA property.

2. Instead of (3.3), it would be sufficient to require that $\int_\Gamma |\log d_z| \, |dz| < \infty$ (Mergelyan [1953, p. 124]). In a certain sense this last condition is best possible (see Mergelyan [1953, p. 158]).

Proof of Theorem 2. Suppose $f \in L^2(G)$, and suppose there exists a sequence of polynomials P_n such that $\|f - P_n\| \to 0$ $(n \to \infty)$. We show that under these conditions f is necessarily analytic in the interior of Γ. Since the last statement does not hold for all $f \in L^2(G)$, the domain G cannot have the PA property.

Lemma 1 in §1 implies that

$$|P_n(z) - P_m(z)| \leq \frac{\|P_n - P_m\|}{\sqrt{\pi}} \cdot \frac{1}{d_z} \quad \text{for each } z \in \Gamma \backslash \{P\}. \tag{3.4}$$

Consequently, by (3.3),

$$\int_\Gamma |P_n(z) - P_m(z)|^\alpha |dz| \leq I \cdot \pi^{-\alpha/2} \|P_n - P_m\|^\alpha.$$

An application of Lemma 2 now shows that the sequence $\{P_n\}$ converges uniformly on each compact subset of the interior of Γ. Since the limit function must coincide with f in the domain G, the function f has an analytic continuation into the interior of Γ.

We make an additional remark about the proof: In the special case where P lies at $z = 1$ and

$$d_z \geq c|z - 1|^p \quad \text{for some } c > 0 \text{ and } p > 0,$$

use of Lemma 2 can be avoided. In this case, inequality (3.4) implies that

$$|(z - 1)^p (P_n(z) - P_m(z))| \leq \frac{1}{c\sqrt{\pi}} \|P_n - P_m\|$$

for each $z \in \Gamma \setminus \{P\}$. The inequality thus holds on all of Γ and therefore also in the interior of Γ. Again we have the uniform convergence of $\{P_n\}$ in the interior of Γ.

Remarks about §3

1. Results are known also for the case where not all polynomials are admitted for the approximation, but only linear combinations of certain powers $z^{\lambda_{n_i}}$; see Mergelyan [1953, p. 146]. For multiply connected domains, polynomials need to be supplemented by rational functions (Mergelyan [1953, p. 114]).

2. If a domain G does not have the PA property, then the class of functions $f \in L^2(G)$ that can be approximated by polynomials is a closed subspace of $L^2(G)$. The problem arises to characterize the elements of this subspace. Havin ([1968a], [1968b]) has studied this for moon-shaped domains.

3. The criterion for polynomials to be dense in $L^2(G)$ that was given in Theorem 1 has been extended to more general sets by Sinanjan [1966]. A compact set $K \subset \mathbb{C}$ is called a *Carathéodory set* if $\partial K = \partial g_\infty$, where g_∞ is the unbounded component of $\mathbb{C} \setminus K$. For $p \geq 1$, set

$$L^p(K) = \left\{ f : \iint_K |f|^p \, dm < \infty, f \text{ analytic in } K^\circ \right\}.$$

Using Mergelyan's theorem, Sinanjan shows that polynomials are dense in $L^p(K)$ if K is a Caratheodory set and $p \geq 1$.

4. Closely related is the problem of approximating functions $f \in L^p(K)$ by *analytic* functions, where K is now an arbitrary compact set. Here the concept of analytic p-capacity is important; see Sinanjan [1966] or the survey article by Melnikov and Sinanjan [1976, p. 731 ff.].

5. Approximation in the pth mean with a weight function $w(z)$ in non-Carathéodory domains has been studied, for example, by Brennan [1973], [1977]. See also the work by Burbea ([1976], [1977], [1978]) on polynomial density in Bers spaces.

Brennan's work [1977] allows discussion of the PA property for inner snakes G. It appears that they may or may not have the PA property, the critical condition being

$$\int_{\partial G} \log(1 - |z|)\, d\mu(z) = -\infty,$$

where μ is the harmonic measure on ∂G with respect to a point $z_0 \in \mathbb{D}\backslash \overline{G}$.

6. Finally, it should be mentioned that the mapping function f of a simply connected domain G onto the unit disk $\mathbb{D}$ can be approximated by polynomials in the uniform norm if and only if (i) G is a Carathéodory domain and (ii) the projections of different prime ends of G are mutually disjoint. See Farrell [1932].

§4. Expansion with respect to ON systems in $L^2(G)$

A function analytic in a disk has a power series expansion. Here we deal with the case where f is analytic in some general domain; but first we review series expansions in Hilbert space.

A. ON expansions in Hilbert space

Suppose H is a Hilbert space and $\{\nu_j\}$ is an ON system in H. For each $x \in H$, we form the Fourier coefficients $\gamma_j = (x, \nu_j)$. The following theorem is well known.

Theorem 1. a) *Minimum property of the Fourier coefficients: The quantity* $\|x - \Sigma_{j=1}^{n} c_j\nu_j\|^2$ *is a minimum if and only if* $c_j = \gamma_j$ $(j = 1, 2, \ldots, n)$.

b) *The minimum in part* a) *equals* $\|x\|^2 - \Sigma_{j=1}^{n} |\gamma_j|^2$.

c) *For each* $x \in H$, *Bessel's inequality holds:* $\Sigma_{j=1}^{\infty} |\gamma_j|^2 \leqslant \|x\|^2$.

Proof. All three assertions are a consequence of the following computations:

$$\begin{aligned}\|x - \Sigma c_j\nu_j\|^2 &= (x - \Sigma c_j\nu_j, x - \Sigma c_j\nu_j)\\ &= \|x\|^2 - \Sigma c_j\overline{\gamma}_j - \Sigma \overline{c}_j\gamma_j + \Sigma|c_j|^2\\ &= \|x\|^2 - \Sigma|\gamma_j|^2 + \Sigma(\gamma_j - c_j)(\overline{\gamma}_j - \overline{c}_j)\\ &= \|x\|^2 + \Sigma|\gamma_j - c_j|^2 - \Sigma|\gamma_j|^2.\end{aligned}$$

Later we shall require that $\{v_j\}$ is a complete ON system (CON system); that is, the linear combinations of the v_j are *dense in* H.

Theorem 2. *The following statements are equivalent.*

a) $\{v_j\}$ *is a CON system.*

b) *For each* $x \in H$, *the relation* $\|x - \sum_{j=1}^n \gamma_j v_j\| \to 0$ $(n \to \infty)$ *holds.*

c) *For each* $x \in H$, *Parseval's identity* $\sum_{j=1}^\infty |\gamma_j|^2 = \|x\|^2$ *holds.*

Proof. The equivalence of statements a) and b) follows from Theorem 1,a). Part b) of Theorem 1 implies that

$$\|x - \textstyle\sum_{j=1}^n \gamma_j v_j\|^2 = \|x\|^2 - \textstyle\sum_{j=1}^n |\gamma_j|^2;$$

hence statements b) and c) are equivalent.

The correspondence $x \mapsto \{\gamma_j\}$, where $\gamma_j = (x, v_j)$ $(j = 1, 2, \ldots)$, provides a mapping from the space H into the space l^2 of all sequences of numbers $\{c_j\}$ with $\Sigma|c_j|^2 < \infty$. The mapping is *onto*, because for each sequence $\{c_j\} \in l^2$ the series $\sum_{j=1}^\infty c_j v_j$ is the limit of the Cauchy sequence $\{\sum_{j=1}^n c_j v_j\}$ and thus is an element of H with Fourier coefficients c_j. If, in addition, the ON system is complete, the mapping is even *one-to-one* as well as onto: If x and y have the same Fourier coefficients, then all Fourier coefficients of $x - y$ are zero, and Theorem 2,c) implies that $x - y = 0$.

B. ON expansions in the space $L^2(G)$

Now suppose $H = L^2(G)$, where, to begin with, G is an arbitrary domain, and suppose $\{\phi_j\}$ is an ON system of functions in $L^2(G)$. For each $f \in L^2(G)$, the Fourier coefficients are

$$\gamma_j = (f, \phi_j) = \iint_G f\overline{\phi_j}\, dm \qquad (j = 1, 2, \ldots),$$

and the *Fourier series* of f becomes

$$f \sim \textstyle\sum_{j=1}^\infty \gamma_j \phi_j.$$

If the ϕ_j form a CON system, Theorem 2 implies that

(4.1) $$\|f - \textstyle\sum_{j=1}^n \gamma_j \phi_j\| \downarrow 0 \qquad (n \to \infty);$$

that is, the Fourier series of f converges to f in the quadratic mean. An even stronger result is true.

Theorem 3. *If $\{\phi_j\}$ is a CON system and $\Sigma_{j=1}^{\infty} \gamma_j \phi_j$ is the corresponding Fourier series of a function $f \in L^2(G)$, then relation* (4.1) *holds and the Fourier series $\Sigma_{j=1}^{\infty} \gamma_j \phi_j$ converges to f uniformly on each compact subset B of G.*

Proof. If $d > 0$ denotes the distance from B to the boundary ∂G, then estimate (1.2) in §1 implies that

$$|f(z) - \Sigma_{j=1}^{n} \gamma_j \phi_j(z)| \leqslant \frac{\|f - \Sigma_1^n \gamma_j \phi_j\|}{\sqrt{\pi}\, d} \qquad (z \in B),$$

and our assertion follows.

As was indicated in the last paragraph of Section A, we can conversely start with a sequence of numbers $\{c_j\} \in l^2$ and form the series $\Sigma c_j \phi_j$. This yields a function $f \in L^2(G)$, where

$$\Sigma_{j=1}^{\infty} c_j \phi_j(z) = f(z) \qquad (z \in G),$$

and the convergence is uniform on compact subsets of G.

Theorem 3 points out the process necessary to obtain a series expansion in G for a function $f \in L^2(G)$:

(i) Suppose $\{u_j\}$ is a system of linearly independent functions in $L^2(G)$;

(ii) one of the ON processes in §2 generates an ON system $\{v_j\}$, which must be complete (§3);

(iii) after computing the Fourier coefficients $\gamma_j = (f, v_j)$, we get an expansion for f in G, which certainly converges in G.

C. The quality of the approximation if f is analytic in $\overline{G}$

Next we deal with the question of when the series expansion of f converges uniformly even in $\overline{G}$. However, we consider only the special case where (i) G is a Jordan domain, that is, $\partial G = C$ is a Jordan curve; (ii) f is analytic in $\overline{G}$; (iii) the functions ϕ_j are the ON polynomials P_j from §2.

It is clear that the ϕ_j must be specially chosen if one wishes to make a statement about the quality of the approximation; for this, even the arrangement of the ϕ_j matters.

As preparation, we need a lemma about polynomials discovered by Bernstein (1912). Let

$$z = \psi(w) = cw + c_0 + \frac{c_1}{w} + \ldots \qquad (c > 0)$$

denote the conformal mapping, normalized at ∞, of the region $\{w: |w| > 1\}$ onto the exterior of C. Suppose ϕ is the inverse mapping, and let

$$C_R = \{z\colon |\phi(z)| = R\} \quad \text{for } R > 1$$

denote a level curve in the exterior of C. With this notation, the following is true.

Bernstein's Lemma. *If P is a polynomial of degree n and $|P(z)| \leq 1$ for $z \in C$, then $|P(z)| \leq R^n$ for $z \in C_R$ (hence also for z in the interior of C_R).*

This lemma describes the growth of polynomials in $\mathbb{C}$. If $C = \{z\colon |z| = 1\}$, then $P(z) = z^n$ shows that the conclusion cannot be strengthened. But if we require, in addition, that $P(z) \neq 0$ ($z \in \mathbb{D}$), one can conclude even that $|P(z)| \leq \frac{1}{2}(R^n + 1)$ for $|z| = R > 1$; see Ankeny and Rivlin [1955].

Proof. The function $P(z)/[\phi(z)]^n$ is analytic in the exterior of C and continuous in the closure of the exterior of C (including ∞). Hence the function must assume its maximum on C:

$$\left|\frac{P(z)}{[\phi(z)]^n}\right| \leq 1 \quad \text{for } z \in \operatorname{ext} C.$$

For $z \in C_R$ we have $|\phi(z)| = R$, and the assertion follows.

We now state our approximation theorem.

Theorem 4. *Suppose C is a Jordan curve and $\rho > 1$ is the largest number such that f is analytic inside C_ρ. Suppose further that $f \sim \sum_{j=1}^{\infty} \gamma_j P_j$ is the expansion of f with respect to the ON polynomials P_j of G, and write $p_n = \sum_{j=1}^{n} \gamma_j P_j$. Then the relation*

$$\max\{|f(z) - p_n(z)| : z \in \overline{G}\} = O(R^{-n}) \tag{4.2}$$

holds for each $R < \rho$, but for no $R > \rho$.

In other words, we have

$$\overline{\lim}_{n\to\infty} \sqrt[n]{\max|f(z) - p_n(z)|} = 1/\rho.$$

Walsh [1969, p. 79] calls this *maximal convergence*; the reason for this will presently become clear.

Proof. a) We show first that there are no polynomials p_n of degree n such that

$$|f(z) - p_n(z)| \leq M_1/R^n \quad \text{for some } R > \rho \text{ and all } z \in \overline{G}. \tag{4.3}$$

If there were such polynomials, we would choose $R_1 \in (\rho, R)$ and find that

$$|p_{n+1}(z) - p_n(z)| \leq |f(z) - p_{n+1}(z)| + |f(z) - p_n(z)| \leq 2M_1/R^n \quad (z \in \overline{G});$$

consequently, by Bernstein's Lemma,

$$|p_{n+1}(z) - p_n(z)| \leqslant \frac{2M_1}{R^n} \cdot R_1^{n+1} = 2M_1 R_1 (R_1/R)^n$$

for $z \in C_{R_1}$. Hence the series of polynomials

$$p_0 + \Sigma_{n=0}^{\infty} (p_{n+1} - p_n)$$

converges uniformly on and inside C_{R_1}, and the limit function F is analytic in int C_{R_1}. But relation (4.3) implies that $F = f$ in G. Hence the function f has an analytic continuation from G to int C_{R_1}, where $R_1 > \rho$. But this contradicts the way ρ was defined.

b) Now we prove the positive part of Theorem 4, that is, statement (4.2) for $R < \rho$. We choose numbers σ and R_1 such that $1 < \sigma < R_1 < \rho$ and show that

$$\max_{\bar{G}} |f(z) - p_n(z)| \leqslant N(\sigma/R_1)^n \qquad (n = 0, 1, 2, \ldots), \tag{4.4}$$

where the p_n are the partial sums of the Fourier series of f and N is some positive constant. Statement (4.2) for $R < \rho$ follows from this.

Here we use the fact that there exist polynomials π_n of degree n such that

$$\max_{\bar{G}} |f(z) - \pi_n(z)| \leqslant M_2 \cdot R_1^{-n} \qquad (n = 0, 1, 2, \ldots);$$

this will be proved in Chapter II, §2 by interpolation. These polynomials satisfy $\|f - \pi_n\| \leqslant M_2 \cdot R_1^{-n}$ in the L^2-norm, and hence the same inequality holds for the minimal polynomials p_n (Theorem 1,a)): $\|f - p_n\| \leqslant M_2 \cdot R_1^{-n}$. Lemma 1 of §1 now implies for each compact subset $B \subset G$ that

$$\max_{B} |f(z) - p_n(z)| \leqslant M_3(B) \cdot R_1^{-n},$$

so that

$$|p_{n+1}(z) - p_n(z)| \leqslant 2M_3(B) \cdot R_1^{-n} \quad \text{for } z \in B.$$

Now we choose $B = \Gamma$, where Γ is a Jordan curve so close to C that C lies in the interior of Γ_σ. Bernstein's Lemma yields

$$|p_{n+1}(z) - p_n(z)| \leqslant 2M_3(B) R_1^{-n} \sigma^{n+1} \quad \text{for } z \in \Gamma_\sigma.$$

Hence the series of polynomials $p_0 + \Sigma_{n=0}^{\infty}(p_{n+1} - p_n)$ converges uniformly on and in the interior of Γ_σ to an analytic function F, and since $\overline{G}$ lies inside Γ_σ, the estimates

$$|F(z) - p_n(z)| = |\Sigma_{k\geqslant n}(p_{k+1}(z) - p_k(z))| \leqslant \Sigma_{k\geqslant n}|p_{k+1}(z) - p_k(z)|$$
$$\leqslant 2M_3(B)\sigma \cdot \Sigma_{k\geqslant n}(\sigma/R_1)^k$$

hold for $z \in \overline{G}$. This establishes (4.4) with F instead of f. But since $p_n(z) \to f(z)$ $(n \to \infty)$ for $z \in G$, the function F is identical with f in G, and the proof of (4.4) is complete.

Statement (4.2) can be generalized. If $|f(z) - p_n(z)| \leqslant M/R^n$ $(z \in \overline{G})$, then $|p_{n+1}(z) - p_n(z)| \leqslant 2M/R^n$ $(z \in C)$, and hence $|p_{n+1}(z) - p_n(z)| \leqslant (2M/R^n)\sigma^{n+1}(z \in C_\sigma)$. This, in turn, implies that

$$\max\ \{|f(z) - p_n(z)|: z \in C_\sigma\} = O((\sigma/R)^n) \qquad (n \to \infty)$$

for all σ and R such that $1 \leqslant \sigma < R < \rho$. Thus $p_n(z)$ converges to $f(z)$ $(n \to \infty)$ even in int C_ρ.

Remarks about §4

1. Even if f is not analytic in $\overline{G}$, it is possible to estimate the error $|f(z) - p_n(z)|$ $(z \in \overline{G})$ between f and the partial sums p_n of the Fourier series of f. However, the proofs are much more difficult; see Suetin [1971, p. 23 ff.]. Suppose, for example, that $C \in C(2, \alpha)$; that is, C has a parametrization $z = z(s)$, where s denotes arc length, and the second derivative of $z(s)$ is continuous and belongs to the class Lip α for some $\alpha > 0$. With this assumption we have

$$|f(z) - p_n(z)| \leqslant \text{const} \cdot \log n \cdot E_n(f, \overline{G}) \qquad (z \in \overline{G};\, n \geqslant 2),$$

where E_n is the minimal error in approximating f by polynomials of degree n.

Estimates of E_n can be obtained through the Faber series of f; see §6,D. In general,

$$E_n(f, \overline{G}) \leqslant \frac{\text{const}}{n^{p+\beta}} \qquad (n \geqslant 1)$$

holds if $f^{(p)}$ is continuous in $\overline{G}$ and $f^{(p)} \in$ Lip β.

2. In the proof of part a) of Theorem 4 we showed that if $|f(z) - p_n(z)| \leqslant M/R^n$ $(z \in \overline{G})$ for polynomials p_n of degree n, then f is analytic in the interior of C_R. In sharp contrast to this is the following result on rational approximation (Aharonov and Walsh [1971], Szüsz [1974]): For each sequence $\{\epsilon_n\}$

converging to zero there exists a function f analytic in $\mathbf{D} = \{z: |z| < 1\}$ having $\partial\mathbf{D}$ as natural boundary, for which $|f(z) - r_n(z)| < \epsilon_n$ $(z \in \mathbf{D})$. Here the r_n are rational functions of order n.

§5. The Bergman kernel function

Now we turn to a special function in $L^2(G)$ and its properties, particularly in relation to the construction of conformal mappings. Appropriate literature was cited preceding §1.

A. Introduction and properties of the kernel function

We choose a functional-analytic introduction and recall the following facts. If H is a Hilbert space and L is a bounded linear functional on H, then there exists a uniquely determined element $u \in H$ such that

$$L(x) = (x, u) \qquad (x \in H).$$

If the Hilbert space is $L^2(G)$ for some arbitrary domain G, Lemma 1 in §1 assures us that $|f(\zeta)| \leqslant \|f\|/(\sqrt{\pi}\, d_\zeta)$, where $d_\zeta = \operatorname{dist}(\zeta, \partial G)$. Hence the functional

$$L(f) := f(\zeta) \qquad (f \in L^2(G))$$

is bounded in $L^2(G)$ for each fixed $\zeta \in G$. Thus there exists a *uniquely determined* $u_\zeta \in L^2(G)$ such that

$$f(\zeta) = (f, u_\zeta) \qquad (f \in L^2(G)).$$

The traditional notation is $u_\zeta(z) =: K(z, \zeta)$, and K is called the *Bergman kernel function* of G. For each $\zeta \in G$, the function K has the *reproducing property*

$$(5.1) \qquad f(\zeta) = (f, K(\cdot, \zeta)) = \iint_G f(z)\overline{K(z, \zeta)}\,dm_z \qquad (f \in L^2(G)).$$

Two properties can be derived from this immediately. a) If one substitutes $f = K(\cdot, \zeta)$ in (5.1), one finds that

$$(5.2) \qquad \|K(\cdot, \zeta)\|^2 = K(\zeta, \zeta) \qquad (\zeta \in G).$$

b) For $z_1, z_2 \in G$, the relation $K(z_1, z_2) = \overline{K(z_2, z_1)}$ holds. To see this, we let $f = K(\cdot, z_2)$ and $\zeta = z_1$ in (5.1), and we obtain

$$K(z_1, z_2) = \iint_G K(z, z_2)\overline{K(z, z_1)}\, dm_z$$

$$= [\iint_G K(z, z_1)\overline{K(z, z_2)}\, dm_z]^- = \overline{K(z_2, z_1)}.$$

The connection between the kernel function and a certain minimum problem in the space $L^2(G)$ is also important. Suppose $\zeta \in G$ is fixed, and write

$$M = \{f \in L^2(G) : f(\zeta) = 1\}.$$

Theorem 1. *There is exactly one solution $f_0 \in M$ such that* $\min_{f \in M} \|f\| = \|f_0\|$. *This function f_0 is connected with the Bergman kernel function as follows:*

$$(5.3) \qquad f_0(z) = \frac{K(z, \zeta)}{K(\zeta, \zeta)} \quad \textit{and} \quad K(z, \zeta) = \frac{f_0(z)}{\|f_0\|^2}.$$

Proof. For each $f \in L^2(G)$, we have $f(\zeta) = (f, K(\cdot, \zeta))$. Using Schwarz's inequality, we find

$$1 = (f, K(\cdot, \zeta)) \leqslant \|f\| \cdot \|K(\cdot, \zeta)\| = \|f\| \cdot \sqrt{K(\zeta, \zeta)} \qquad (f \in M).$$

Equality occurs if and only if $f = f_0 = CK(\cdot, \zeta)$. Because of the relation $1 = f_0(\zeta) = CK(\zeta, \zeta)$, we see that

$$f_0(z) = K(z, \zeta)/K(\zeta, \zeta).$$

The second statement in (5.3) also follows from the equations above:

$$K(z, \zeta) = f_0(z)K(\zeta, \zeta) = f_0(z)/\|f_0\|^2.$$

Often the relations expressed in Theorem 1 are used to define the Bergman kernel function.

B. Series representation of the Bergman kernel function

In only a few cases is it possible to get a representation in closed form for the kernel function. However, it is easy to find a series expansion with respect to some CON system $\{\phi_j\}$, because, by (5.1), the Fourier coefficients are simply

$$\gamma_j = (K(\cdot, \zeta), \phi_j) = \overline{\phi_j(\zeta)} \qquad (j = 1, 2, \ldots).$$

Using Theorem 3 in §4, we therefore obtain the following result.

Theorem 2. *If $\{\phi_j\}$ denotes an arbitrary CON system, the Bergman kernel function has the representation*

$$K(z, \zeta) = \Sigma_{j=1}^{\infty} \overline{\phi_j(\zeta)}\phi_j(z) \qquad (z, \zeta \in G). \tag{5.4}$$

For each fixed $\zeta \in G$, the series converges uniformly on each compact subset B of G.

This theorem allows the possibility of an actual approximation of the kernel function. Theorem 4 in §4 deals with the rate of convergence of the series (5.4) in $\overline{G}$ (where G is a Jordan domain) in the case where $K(\cdot, \zeta)$ is analytic in $\overline{G}$ and the functions ϕ_j are the ON polynomials P_j of G. Examples demonstrate, however, that it is often desirable to augment the system $\{P_j\}$ with additional functions in order to speed up the convergence; see Section C_3.

Now we consider the *special case* where $G = \mathbb{D} = \{z\colon |z| < 1\}$. According to §2,A we have

$$P_n(z) = \sqrt{\frac{n+1}{\pi}}\, z^n \qquad (n = 0, 1, 2, \ldots).$$

Thus

$$K(z, \zeta) = \Sigma_{n=0}^{\infty} \frac{n+1}{\pi} \overline{\zeta}^n z^n = \frac{1}{\pi} \cdot \frac{1}{(1 - z\overline{\zeta})^2} \qquad (z, \zeta \in \mathbb{D})$$

is the kernel function of $\mathbf{D}$. The series converges in $\overline{\mathbf{D}}$, but the rate of convergence deteriorates more and more as ζ approaches $\partial\mathbf{D}$. The reproducing property (5.1) becomes

$$f(\zeta) = \frac{1}{\pi} \iint_{\mathbb{D}} \frac{f(z)}{(1 - \overline{z}\zeta)^2}\, dm_z \qquad (\zeta \in \mathbf{D})$$

and is valid for each $f \in L^2(\mathbf{D})$.

If ∂G is an ellipse, the ON polynomials can be expressed in terms of Chebyshev polynomials, and the series (5.4) of K can also be given explicitly; see Nehari [1952, pp. 258-259].

C. Construction of conformal mappings with the Bergman kernel function

The Bergman kernel function is an important tool for the numerical construction of conformal mappings. Here we shall limit ourselves to simply connected domains. In the case of multiply connected domains, Nehari [1952, pp. 367-377] and Gaier [1964, Chapter V, §5] use the kernel function to construct conformal mappings onto canonical domains.

C_1. The connection between K and conformal mapping

Suppose $G \subset \mathbb{C}$ ($G \neq \mathbb{C}$) is a simply connected domain, ζ is a fixed point in G, and F is the conformal mapping of G onto $\mathbb{D}$, normalized by the conditions $F(\zeta) = 0$ and $F'(\zeta) > 0$. As is well known, these conditions determine F uniquely.

Theorem 3. *The conformal mapping F and the Bergman kernel function K of G are related as follows:*

$$(5.5)\quad F'(z) = \sqrt{\frac{\pi}{K(\zeta,\zeta)}}\, K(z,\zeta) \ \text{and}\ K(z,\zeta) = \frac{1}{\pi} F'(z)\overline{F'(\zeta)} \ \text{for } z \in G.$$

Proof. We show that $F'(z)\overline{F'(\zeta)}/\pi$ has the reproducing property and therefore is identical to $K(z, \zeta)$. To this end, suppose $f \in L^2(G)$, and write $G_\rho := \{z: |F(z)| < \rho\}$ $(0 < \rho < 1)$. Since $\overline{F(z)}\cdot F(z) = \rho^2$ for $z \in \partial G_\rho$, Green's formula implies

$$\iint_{G_\rho} f\overline{F'}\, dm_z = \frac{1}{2i}\int_{\partial G_\rho} f\overline{F}\, dz = \frac{\rho^2}{2i}\int_{\partial G_\rho} \frac{f}{F}\, dz.$$

The last integral can be evaluated using the residue theorem, and it equals $2\pi i f(\zeta)/F'(\zeta)$. Hence

$$f(\zeta) = \frac{1}{\rho^2}\iint_{G_\rho} f(z)\, \frac{\overline{F'(z)\overline{F'(\zeta)}}}{\pi}\, dm_z.$$

For $\rho \to 1$ we obtain (5.1) with $K(z,\zeta) = F'(z)\overline{F'(\zeta)}/\pi$. For $z = \zeta$ it follows that $K(\zeta,\zeta) = (F'(\zeta))^2/\pi$, and the first equation in (5.5) is also established.

We observe that K can also be expressed in terms of a *nonnormalized* conformal mappoing F of G onto $\mathbb{D}$. Inserting an auxiliary linear transformation, one finds easily that

$$K(z,\zeta) = \frac{1}{\pi}\cdot\frac{F'(z)\,\overline{F'(\zeta)}}{[1 - \overline{F(\zeta)}F(z)]^2} \qquad (z,\zeta \in G).$$

In the special case where $G = \mathbb{D}$, we can choose $F(z) = z$ and obtain again the kernel function of $\mathbb{D}$.

To summarize: Using the series representation (5.4) of K, we can express the conformal mapping F (normalized such that $F(\zeta) = 0$ and $F'(\zeta) > 0$) of G onto $\mathbb{D}$ as

$$(5.6) \qquad F(z) = \sqrt{\frac{\pi}{K(\zeta,\zeta)}} \int_{\nu=\zeta}^{z} K(\nu,\zeta)\, d\nu \qquad (z \in G).$$

If f is the conformal mapping of G onto the disk $\{w: |w| < r\}$, normalized such that $f(\zeta) = 0$ and $f'(\zeta) = 1$, we get $r = (\pi K(\zeta,\zeta))^{-1/2}$ and

$$(5.7) \qquad f(z) = \frac{F(z)}{F'(\zeta)} = \frac{1}{K(\zeta,\zeta)} \int_{\nu=\zeta}^{z} K(\nu,\zeta)\, d\nu \qquad (z \in G).$$

C_2. The Bieberbach polynomials

In practice, the series (5.4) has to be truncated and will therefore yield only approximations for F and f, respectively. If G is a Jordan domain and the ϕ_j are the ON polynomials P_j of G, we obtain approximating polynomials

$$\pi_n(z) := \frac{1}{K_{n-1}(\zeta,\zeta)} \int_{\nu=\zeta}^{z} K_{n-1}(\nu,\zeta)\, d\nu,$$

where

$$K_{n-1}(z,\zeta) = \Sigma_{j=0}^{n-1} \overline{P_j(\zeta)} P_j(z) \qquad (n = 1, 2, \dots).$$

These π_n are called the *Bieberbach polynomials* of G and ζ. They satisfy the conditions

$$\pi_n(\zeta) = 0, \qquad \pi_n'(\zeta) = 1, \qquad \|\pi_n'\|^2 = \frac{1}{K_{n-1}(\zeta,\zeta)},$$

and it is easy to see that $\|\pi_n'\| = \min \|p'\|$, where the minimum is taken over all polynomials p of degree n such that $p(\zeta) = 0$ and $p'(\zeta) = 1$.

Concerning the convergence of the sequence $\{\pi_n\}$, Theorem 2 and (5.7) imply that $\pi_n(z) \Rightarrow f(z)$ $(n \to \infty)$ on each compact subset B of G. Further, the inequality

$$|f(z) - \pi_n(z)| \leqslant M q^n \qquad (z \in \overline{G}; n = 1, 2, \dots)$$

holds for some $q < 1$ if and only if f is analytic in $\overline{G}$. The latter is the case if ∂G is an analytic Jordan curve, but also if ∂G is the circumference of a square, for example.

Weaker assumptions about ∂G require more detailed investigations. Gaier [1964, p. 125] reports on earlier results by Keldysh and Mergelyan. The present state of the art is as follows. We say that the boundary C of G belongs to the class $C(p, \alpha)$ if C is rectifiable and has a representation $z = z(s)$ in terms of the arc length s with a continuous p^{th} derivative in the class Lip α ($p = 1, 2, \ldots; 0 < \alpha \leqslant 1$). With this notation, the following theorem is true.

Theorem 4. a) *Suppose $C \in C(p, \alpha)$ and $p + \alpha \geqslant 7/4$. Then the Bieberbach polynomials π_n satisfy*

$$|f(z) - \pi_n(z)| \leqslant \text{const } n^{-p-\alpha} \cdot \log n \qquad (z \in \overline{G}).$$

b) *If $C \in C(1, \alpha)$ and $1/2 < \alpha < 3/4$, then*

$$|f(z) - \pi_n(z)| \leqslant \text{const } n^{-3\alpha + 1/2} \qquad (z \in \overline{G}).$$

c) *If $C \in C(1, \alpha)$ and $\alpha > 0$, then*

$$|f(z) - \pi_n(z)| \leqslant \text{const } n^{-\alpha - 1/2} \cdot \sqrt{\log n} \qquad (z \in \overline{G}).$$

d) *If the boundary C of G is smooth and of bounded curvature, then*

$$|f(z) - \pi_n(z)| \leqslant \text{const } (\log n)^2/n^2 \qquad (z \in \overline{G}).$$

All of these results are written up in detail by Suetin [1971, Chapter V]. There are also estimates for $z \in B \subset G$, which depend on the quality of ∂G. The proofs depend on an asymptotic representation of the ON polynomials P_n in $\overline{G}$ and the best-approximation polynomials of degree n corresponding to f.

Probably the first paper giving estimates for regions G with *piecewise* smooth boundary C is that of Simonenko [1978]. Here G is assumed to be a Lipschitz domain, and in particular, C may be a polygon. For any such domain there exist constants $c > 0$ and $\gamma > 0$ such that

$$|f(z) - \pi_n(z)| \leqslant c/n^\gamma \qquad (z \in \overline{G}).$$

Kulikov [1979] permits more general boundaries, but the estimate is only in the $L^p(G)$ norm.

C_3. The use of singular functions in the ON process

The rate of convergence of the series (5.4) often is poor if one chooses for the ϕ_j simply the ON polynomials P_j of G. The reason for this stems from the singularities of the mapping function F, which may lie on ∂G or in $\mathbb{C}\backslash\overline{G}$ but close to ∂G. However, the rate of convergence of the series (5.4) can often be much improved if one supplements the powers z^k $(k = 0, 1, 2, \ldots)$ by some additional functions whose singularities reflect those of K. Only the augmented system will then undergo the ON process. Levin, Papamichael, and Sideridis [1978] were the first to suggest this, and they have thus considerably increased the applicability of the Bergman kernel function to conformal mappings.

If, for example, G is the rectangle $\{z = x + iy\colon |x| < a, |y| < 1\}$ and $\zeta = 0$, and if F is the mapping of G onto $\mathbf{D}$, normalized such that $F(0) = 0$ and $F'(0) > 0$, then the reflection principle shows that the poles of the analytic continuation of F that are closest to ∂G lie at the points $z = \pm 2a, \pm 2i$ and are simple. In a series representation of F these poles can be taken into account by terms of the form $z/(z^2 - 4a^2)$ and $z/(z^2 + 4)$. For $K =$ const F', this means that the basis $\{z^k\}_{k=0}^{\infty}$ should be enlarged by the terms

$$\frac{d}{dz}\left(\frac{z}{z^2 - 4a^2}\right) \quad \text{and} \quad \frac{d}{dz}\left(\frac{z}{z^2 + 4}\right),$$

and the enlarged system should then undergo the ON process.

Suppose the ON process yields the functions ϕ_j. If we write (note that this differs from Section C_2)

$$K_n(z, \zeta) := \Sigma_{j=1}^{n} \overline{\phi_j(\zeta)}\phi_j(z),$$

then

$$F_n(z) := \sqrt{\frac{\pi}{K_n(\zeta, \zeta)}} \int_{v=\zeta}^{z} K_n(v, \zeta)\, dv \qquad (z \in G)$$

is an approximation of the conformal mapping F of G onto $\mathbf{D}$.

In the work cited above, Levin, Papamichael, and Sideridis [1978] carry out the computations for a rectangle with $a = 2$, for example. If the ON process is applied to the monomials z^k, then 17 functions are necessary in order to achieve an approximation error of $2 \cdot 10^{-5}$. But if the two singular functions $\left(\frac{z}{z^2 - 16}\right)'$ and $\left(\frac{z}{z^2 + 4}\right)'$ as well as powers z^k are used in the ON process, then a total of 10 functions will reduce the approximation error to only $2 \cdot 10^{-10}$. The use of this method is also advisable if F has singularities on ∂G, as would be the case with conformal mappings of polygons. It only has a certain disadvantage in that the inner products that occur during the orthonormalization are somewhat more difficult to evaluate if singular functions are included in the process.

D. Additional applications of the Bergman kernel function

Finally we mention two further interesting applications of the Bergman kernel function in complex analysis.

D_1. Domains with the mean-value property

First, suppose G is the disk $\{z: |z - \zeta| < r\}$ and f is analytic in G. It is well known that

$$f(\zeta) = \frac{1}{2\pi} \int_0^{2\pi} f(\zeta + \rho e^{i\phi})\, d\phi \qquad (0 < \rho < r).$$

Multiplying by ρ and then integrating from 0 to r, we see that

$$f(\zeta) = \frac{1}{A(G)} \iint_G f(z)\, dm_z \quad \text{for each} \quad f \in L^1(G), \tag{5.8}$$

where $A(G) = \pi r^2$ denotes the area of G. We now show that the converse also holds.

Theorem 5. *Suppose G is a simply connected, bounded domain with area $A(G)$, and suppose ζ is a fixed point in G. Then* (5.8) *holds for all functions f that are analytic and integrable in G if and only if G is a disk with center ζ.*

Proof. We apply (5.8) only to functions $f \in L^2(G)$. (By Schwarz's inequality, $L^2(G) \subset L^1(G)$.) If we write (5.8) in the form

$$f(\zeta) = \iint_G f(z)\, \frac{1}{A}\, dm_z,$$

we see that $1/A$ has the reproducing property. Therefore $K(z, \zeta) = 1/A$, and by Theorem 3 we have $F'(z) = \sqrt{\pi/A}$ for the normalized (in ζ) conformal mapping F of G onto $\mathbb{D}$. Hence $F(z) = \text{const} \cdot (z - \zeta)$, and the assertion follows.

D_2. Representation of $\int_{-1}^{+1} f(x)dx$ as an area integral

In Section A we used the functional $L(f) = f(\zeta)$ (for fixed $\zeta \in G$) as a starting point in order to introduce the Bergman kernel function. Now suppose G is a domain containing the interval $I = [-1, +1]$, and consider the linear functional

$$L(f) := \int_I f(x)dx \qquad (f \in L^2(G)).$$

If $d = \operatorname{dist}(I, \partial G)$, Lemma 1 in §1 implies that

$$|L(f)| \leqslant \frac{2}{\sqrt{\pi}\, d} \|f\|.$$

Hence our functional is bounded, and there exists a uniquely determined element $h \in L^2(G)$ such that

$$\int_I f(x)dx = (f, h) \qquad (f \in L^2(G)).$$

This function h can be expressed in terms of the kernel function. For $z \in G$ we see that

$$\begin{aligned} \int_I K(z, x)dx &= [\int_I K(x, z)dx]^- = [(K(\cdot, z), h)]^- \\ &= (h, K(\cdot, z)) = h(z), \end{aligned}$$

where the last equality is due to the reproducing property (5.1) of K. This would already express $\int_I f\,dx$ as an area integral, but we ask more specifically: *Is there a domain $G \supset I$ such that*

$$\int_I f(x)\,dx = \iint_G f(z)\,dm_z \quad \textit{for each } f \in L^2(G)?$$

Davis [1969] posed and dealt with this question. According to the above, we must have $h = 1$, and the question now becomes: *Is there a domain $G \supset I$ whose kernel function K satisfies the condition*

$$(5.9) \qquad \int_I K(z, x)dx = 1 \qquad (z \in G)?$$

Suppose now that G is simply connected and symmetric with respect to the real axis $\mathbf{R}$. Consider the conformal mapping F of G onto $\mathbf{D}$ that maps $[-1, +1]$ onto an interval $[-\alpha, +\alpha] \subset \mathbf{D}$. Then the kernel function is given by the expression (see Section C_1)

$$K(z, x) = \frac{1}{\pi} \cdot \frac{F'(z)F'(x)}{(1 - F(x)F(z))^2} \qquad (z \in G, x \in I),$$

and condition (5.9) becomes

$$\int_I \frac{F'(z)F'(x)}{(1 - F(x)F(z))^2}\, dx = \pi \qquad (z \in G),$$

or, equivalently,

$$\int_I \frac{d}{dx}\left(\frac{1}{1 - F(x)F(z)}\right) dx = \pi\, \frac{F(z)}{F'(z)} \qquad (z \in G).$$

We integrate the left-hand side and obtain

$$F'(z) = \frac{\pi}{2\alpha}\,(1 - \alpha^2 F(z)^2) \qquad (z \in G).$$

Integration of this differential equation leads to the relations

(5.10) $$z = \frac{1}{\pi} \log \frac{1 + \alpha w}{1 - \alpha w}, \qquad w = \frac{1}{\alpha}\, \frac{e^{\pi z} - 1}{e^{\pi z} + 1},$$

where $w = F(z)$. Since $F(1) = \alpha$, it follows that

$$\alpha = \sqrt{\frac{e^{\pi} - 1}{e^{\pi} + 1}} = 0.95768 \ldots .$$

Thus we have proved the following *result*: The image of $\mathbf{D}$ under the mapping

$$w \mapsto \frac{1}{\pi} \log \frac{1 + \alpha w}{1 - \alpha w}, \quad \text{where } \alpha = \sqrt{\frac{e^{\pi} - 1}{e^{\pi} + 1}},$$

is a domain G with the property that

$$\int_{-1}^{+1} f(x)\, dx = \iint_G f(z)\, dm_z \quad \text{for each } f \in L^2(G).$$

One finds that ∂G is a convex, ellipse-shaped Jordan curve through the points $(\pm 1.2205, 0)$ and $(0, \pm 0.4862)$; see the drawing below.

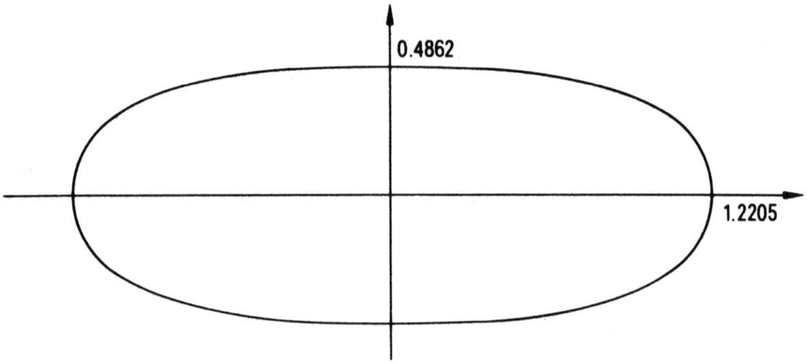

The following *corollary* is also interesting. It is well known that the normalized Legendre polynomials satisfy the orthogonality relations

$$\int_I P_n(x)P_m(x)dx = \delta_{nm}.$$

Hence the non-Hermitian orthogonality relations

$$\iint_G P_n(z)P_m(z)dm_z = \delta_{nm}$$

are valid for the domain G constructed above.

Remark about §5

In Section C we already mentioned that ON expansions can be used also for conformal mappings of multiply connected domains. If G is a ring domain, one is particularly interested in the conformal modulus M of G. Here, also, the use of suitable singular functions can provide a better approximation of M with fewer terms. For this topic see Eidel [1979].

§6. The quality of the approximation; Faber expansions

Suppose the function f is analytic in a domain G and continuous in $\overline{G}$. How accurately such an f can be approximated by polynomials of degree n obviously depends on the properties of both f and G. We have already touched on this topic in several special cases. Now we show how results of this type can be obtained by expanding f in terms of the Faber polynomials of G.

These expansions are less constructive than the ON expansions considered so far, because a generally unknown mapping function enters into the considerations.

A. Boundary behavior of Cauchy integrals

Cauchy integrals are integrals of the form

$$\frac{1}{2\pi i} \int_C \frac{h(\zeta)}{\zeta - z} d\zeta ,$$

where C is a rectifiable Jordan curve, h is integrable on C, and $z \notin C$. For our purposes it is sufficient to consider the special case where

$$\Phi(z) := \frac{1}{2\pi i} \int_{|\zeta|=1} \frac{h(\zeta)}{\zeta - z} d\zeta, \tag{6.1}$$

in other words, where $C = \{\zeta : |\zeta| = 1\}$. We also assume that the function h is *continuous on C*. Relation (6.1) defines two analytic functions, one in the disk $\mathbb{D} = \{z : |z| < 1\}$ and one in its exterior $\{z : |z| > 1\}$.

We ask first how Φ is related to h and its (Fourier) conjugate $\widetilde{h}$.

Lemma 1. *Suppose H and $\widetilde{H}$ denote the harmonic extensions to $\mathbb{D}$, obtained through the Poisson integral, of h and $\widetilde{h}$, respectively. Then*

$$\Phi(z) = \frac{1}{2}\Phi(0) + \frac{1}{2}[H(z) + i\widetilde{H}(z)] \qquad (z \in \mathbb{D}). \tag{6.2}$$

If in addition to h, the conjugate $\widetilde{h}$ is also continuous on $\partial\mathbb{D}$, then Φ has a continuous extension to $\overline{\mathbb{D}}$ and

$$\Phi(e^{i\phi}) = \frac{1}{2}\Phi(0) + \frac{1}{2}[h(e^{i\phi}) + i\widetilde{h}(e^{i\phi})] . \tag{6.3}$$

Proof. We have (see, for example, Duren [1970, p. 62])

$$H(z) + i\widetilde{H}(z) = \frac{1}{2\pi}\int_0^{2\pi} \frac{e^{it} + z}{e^{it} - z} h(e^{it})dt \qquad (z \in \mathbb{D}).$$

On the other hand,

$$2\Phi(z) - \Phi(0) = \frac{1}{\pi}\int_0^{2\pi} h(e^{it}) \left[\frac{e^{it}}{e^{it} - z} - \frac{1}{2}\right] dt$$

$$= \frac{1}{2\pi}\int_0^{2\pi} h(e^{it}) \frac{e^{it} + z}{e^{it} - z} dt.$$

This establishes (6.2). If both h and $\tilde{h}$ are continuous, it is well known that $H(z) \to h(e^{i\phi})$ and $\tilde{H}(z) \to \tilde{h}(e^{i\phi})$ for $z \to e^{i\phi}$, and (6.3) follows.

What can be said if the part of Φ defined in $\mathbb{D}$ vanishes identically?

Lemma 2. a) *The function* Φ *defined by* (6.1) *is analytic in* $\{z: |z| > 1\}$, *and* $\Phi(\infty) = 0$.

b) *If* h *is continuous on* $\partial\mathbb{D}$ *and if* $\Phi(z) = 0$ *for each* $z \in \mathbb{D}$, *then* Φ *has a continuous extension from* $\{z: |z| > 1\}$ *to* $\{z: |z| \geqslant 1\}$; *moreover,*

$$\Phi(z) = -h(z) \quad \text{for} \quad |z| = 1. \tag{6.4}$$

Proof. Part a) is obvious. For part b), we consider the relation

$$\Phi(z) - \Phi\left(\frac{1}{\bar{z}}\right) = \frac{1}{2\pi}\int_0^{2\pi} h(e^{it})P_z(t)dt \qquad (z \in \mathbb{D}),$$

where P_z denotes the Poisson kernel. If $z \to e^{i\phi}$ $(z \in \mathbb{D})$, the right-hand side approaches $h(e^{i\phi})$, and the assertion follows.

General properties of Cauchy integrals can be found in the book by Privalov [1956, pp. 136 ff.]. For recent results on the continuity of Cauchy integrals, see Dynkin [1980].

B. Faber polynomials and Faber expansions

The polynomials introduced by Faber [1903] play an important role in many branches of complex analysis. Here we shall provide only the properties necessary for our approximation theorems. There are survey articles by Suetin [1964], [1976] and Curtiss [1971]; see also Smirnov and Lebedev [1968, Chapter 2]. Important properties of Faber polynomials can be found in the works by Pommerenke [1964], [1965] and Kövari and Pommerenke [1967].

Suppose C is a Jordan curve in the z-plane, and let

$$z = \psi(w) = cw + c_0 + \frac{c_1}{w} + \dots \qquad (c > 0)$$

denote the conformal mapping, normalized at ∞, of $\{w: |w| > 1\}$ onto ext C. Suppose ϕ denotes the inverse mapping:

$$w = \phi(z) = dz + d_0 + \frac{d_1}{z} + \dots \qquad (cd = 1).$$

One possible way of defining Faber polynomials involves the Laurent series

$$[\phi(z)]^n = d^n z^n + \Sigma_{k=-\infty}^{n-1} d_{nk} z^k,$$

which is valid for large $|z|$; its polynomial part is called the nth *Faber polynomial* F_n $(n = 0, 1, 2, \ldots)$. The term of highest degree is $d^n z^n$, so that F_n is exactly of degree n. In special cases the F_n can be given explicitly (see Curtiss [1971, p. 582]).

By definition,

$$[\phi(z)]^n = F_n(z) + H_n(z), \tag{6.5}$$

where H_n is analytic in ext C and $H_n(z) = O(1/|z|)$ for $z \to \infty$. If we change back to w, we see that

$$\begin{aligned} F_n(\psi(w)) &= w^n - H_n(\psi(w)) \\ &= w^n + \Sigma_{k=1}^{\infty} n b_{nk} w^{-k} \qquad (|w| > 1). \end{aligned} \tag{6.6}$$

The b_{nk} are the *Grunsky coefficients*; however, we shall not pursue this (see Pommerenke [1975, Chapter 3]). Relation (6.6) leads to the following result.

Lemma 3. *For nonnegative values of m and n we have*

$$\frac{1}{2\pi i} \int\limits_{|w|=1} \frac{F_m(\psi(w))}{w^{n+1}}\, dw = \begin{cases} 1 & \text{if } m = n, \\ 0 & \text{if } m \neq n. \end{cases} \tag{6.7}$$

Proof. In (6.7) we can integrate over $\{w\colon |w| = R > 1\}$. If we split F_m into w^m and the series according to (6.6), we see that the integral corresponding to the series vanishes, leaving

$$\frac{1}{2\pi i} \int\limits_{|w|=R} w^{m-n-1} dw.$$

This establishes (6.7).

Next we shall find a generating function for the Faber polynomials F_n. Suppose $C_R = \{z\colon |\phi(z)| = R > 1\}$ is a level curve in the exterior of C, and suppose $z \in \operatorname{int} C_R$. Equation (6.5) implies that

$$\frac{1}{2\pi} \int\limits_{C_R} \frac{(\phi(\zeta))^n}{\zeta - z}\, d\zeta = \frac{1}{2\pi i} \int\limits_{C_R} \frac{F_n(\zeta)}{\zeta - z}\, d\zeta + \frac{1}{2\pi i} \int\limits_{C_R} \frac{H_n(\zeta)}{\zeta - z}\, d\zeta.$$

The first integral on the right-hand side equals $F_n(z)$ while the second integral vanishes, because $H_n(\zeta) = O(1/|\zeta|)$ for $|\zeta| \to \infty$. For $z \in \operatorname{int} C_R$ $(R > 1)$, we thus see that

$$(6.8) \qquad F_n(z) = \frac{1}{2\pi i}\int_{C_R} \frac{(\phi(\zeta))^n}{\zeta - z}\, d\zeta = \frac{1}{2\pi i}\int_{|w|=R} \frac{w^n \psi'(w)}{\psi(w) - z}\, dw.$$

If we now write

$$(6.9) \qquad \frac{w\psi'(w)}{\psi(w) - z} = 1 + p_1(z)/w + p_2(z)/w^2 + \ldots \quad (|w| = R,\ z \in \operatorname{int} C_R)$$

and integrate term by term, we find that $F_n(z) = p_n(z)$. Hence (6.9) is a generating function for the Faber polynomials F_n.

Finally, the F_n can also be obtained recursively (Curtiss [1971, p. 579]). But this has little practical significance, since the coefficients in the recursion formula contain the generally unknown numbers c_k.

Now let $G = \operatorname{int} C$ and write

$$A(\overline{G}) = \{F{:}F \text{ analytic in } G \text{ and continuous in } \overline{G}\}.$$

We define the *Faber coefficients* of a function $F \in A(\overline{G})$ by

$$(6.10) \qquad a_n = \frac{1}{2\pi i}\int_{|w|=1} F(\psi(w))w^{-n-1}dw \qquad (n = 0, 1, 2, \ldots),$$

and we call

$$F(z) \sim \Sigma_{n=0}^{\infty} a_n F_n(z)$$

the formal *Faber series of F*. Lemma 3 makes the following statement obvious: *If* $F(z) := \Sigma_{n=0}^{\infty} a_n F_n(z)$ *and if the convergence is uniform in* $\overline{G}$, *then the* a_n *are the Faber coefficents of F.*

In order to identify functions in $A(\overline{G})$, the following theorem is important.

Theorem 1. *Suppose C is a rectifiable Jordan curve,* $G = \operatorname{int} C$, *and* $F \in A(\overline{G})$. *If all Faber coefficients of F vanish, then* $F = 0$.

Proof. From $\frac{1}{2\pi i}\int_{|\omega|=1} F(\psi(\omega))\omega^{-n-1}d\omega = 0$ $(n = 0, 1, 2, \ldots)$ it follows that

$$\sum_{n=0}^{\infty} \frac{1}{2\pi i}\int_{|\omega|=1} F(\psi(\omega))(w/\omega)^{n+1}\, d\omega = w \cdot \frac{1}{2\pi i}\int_{|\omega|=1} \frac{F(\psi(\omega))}{\omega - w}\, d\omega = 0$$

for $w \in \mathbb{D}$. Hence the Cauchy integral corresponding to the continuous function $F \circ \psi$ vanishes in $\mathbb{D}$. Thus Lemma 2 guarantees the existence of a function H continuous in $\{w: |w| \geqslant 1\}$, analytic in $\{w: |w| > 1\}$, and such that

$$H(\infty) = 0 \quad \text{and} \quad H(w) = (F \circ \psi)(w) \quad \text{if} \quad |w| = 1.$$

We define

$$g(z) = \begin{cases} F(z) & \text{if } z \in \overline{G}, \\ H(\phi(z)) & \text{if } z \notin \overline{G}. \end{cases}$$

This function g is analytic in int C and in ext C, and on the curve C the function has the common boundary values $F(z)$. Hence g is continuous in all of $\mathbb{C}$. Since C is rectifiable, the principle of continuity applies, and we conclude that g is an entire function. Because $g(\infty) = 0$, we see that $g = 0$ in $\mathbb{C}$, and consequently $F = 0$.

C. The Faber mapping as a bounded operator

In order to use Faber polynomials for the approximation of a function $F \in A(\overline{G})$, more must be assumed about the boundary $C = \partial G$.

C_1. Curves of bounded rotation

The notion of a curve of bounded rotation was first introduced by Radon [1919]. His goal was to apply the method of integral equations to the solution of the Dirichlet problem for domains with corners. Suppose C is a rectifiable Jordan curve, and let L denote its length. Then the angle $\theta(s)$ of the tangent exists for almost all $s \in (0, L)$.

Definition 1. *The curve C is of bounded rotation if $\theta(s)$ has a continuation to $[0, L]$ such that the extended function is of bounded variation.*

If C is of bounded rotation, we say that $C \in BR$. In order for C to belong to the class BR it is sufficient, for example, that C is made up of finitely many convex arcs (corners are permitted). If $C \in BR$, then there are two half-tangents at each point of C. Further, the inequality

$$\int_C |d_\zeta \arg(\zeta - z)| \leqslant \int_C |d\theta(s)| =: V \tag{6.11}$$

holds for each point $z \in C$ (Radon [1919, p. 1133]). Here the jump of $\arg(\zeta - z)$ at the point z is equal to the exterior angle of C at z. The quantity V is called the *total rotation of C.*

For our purposes it is important that the mapping $z = \psi(w)$ of $\{w\colon |w| > 1\}$ onto ext C can be expressed explicitly in terms of the function

$$\arg(\zeta - z) = \arg(\psi(e^{it}) - \psi(e^{i\theta})) =: \nu(t, \theta);$$

see Paatero [1933] and Pommerenke [1965, p. 425]. For if $c = \psi'(\infty)$ denotes the capacity of C, then

$$\text{(6.12)} \qquad \log \frac{\psi(w) - \psi(e^{i\theta})}{cw} = \frac{1}{\pi} \int_0^{2\pi} \log \left(1 - \frac{e^{it}}{w}\right) d_t\nu(t, \theta) \qquad (|w| > 1).$$

The connection with the Faber polynomials arises if we differentiate (6.12) with respect to w and then multiply by w:

$$\frac{w\psi'(w)}{\psi(w) - \psi(e^{i\theta})} - 1 = \frac{1}{\pi} \int_0^{2\pi} \Sigma_{n=1}^{\infty} \left(\frac{e^{it}}{w}\right)^n d_t\nu(t, \theta)$$

$$= \Sigma_{n=1}^{\infty} w^{-n} \cdot \frac{1}{\pi} \int_0^{2\pi} e^{int}\, d_t\nu(t, \theta).$$

If we compare this with (6.9), we find the fundamental relation

$$\text{(6.13)} \qquad \frac{1}{\pi} \int_0^{2\pi} e^{int}\, d_t\nu(t, \theta) = F_n(\psi(e^{i\theta})) \qquad (n = 1, 2, \ldots);$$

see Pommerenke [1965, p. 425]. Inequality (6.11) implies that for each fixed θ the variation of $\nu(\cdot, \theta)$ is at most V:

$$\text{(6.14)} \qquad \int_0^{2\pi} |d_t\nu(t, \theta)| \leqslant V.$$

Relation (6.13) can be derived more directly (Ellacott). For fixed $z \in \partial G$, we have from (6.8)

$$F_n(z) = \frac{1}{2\pi i} \int_{|w|=R} \frac{w^n \psi'(w)}{\psi(w) - z}\, dw \qquad (n \geqslant 1),$$

and it is obvious that

$$0 = \frac{1}{2\pi i} \int_{|w|=R} \frac{w^{-n} \psi'(w)}{\psi(w) - z}\, dw,$$

since the integrand is $O(|w|^{-2})$ for large $|w|$. This gives us

$$F_n(z) = \frac{1}{2\pi i} \int_{|w|=R} w^n d_w \log[\psi(w) - z]$$

and $0 = \frac{1}{2\pi i} \int_{|w|=R} \bar{w}^n d_w \log[\psi(w) - z]$ or $0 = \frac{1}{2\pi i} \int_{|w|=R} w^n d_w \overline{\log[\psi(w) - z]}$.

Subtracting the last relation from $F_n(z)$, we obtain

$$F_n(z) = \frac{1}{\pi} \int_{|w|=R} w^n d_w \arg[\psi(w) - z] = \frac{R^n}{\pi} \int_0^{2\pi} e^{int}\, d_t \arg[\psi(Re^{it}) - z].$$

Now we wish to let $R \to 1$. To do this, we first integrate by parts to get

$$F_n(z) = 2R^n - \frac{R^n}{\pi} \int_0^{2\pi} \arg[\psi(Re^{it}) - z]\; d(e^{int}).$$

Here we *can* let $R \to 1$ since we have (6.11) and therefore $\arg[\psi(Re^{it}) - z]$ is bounded in t, uniformly with respect to R. After letting $R \to 1$, we integrate by parts again, and

$$F_n(z) = \frac{1}{\pi} \int_0^{2\pi} e^{int}\, d_t \arg[\psi(e^{it}) - z]$$

follows.

C_2. The Faber mapping T

We consider the mapping

$$T\colon w^n \mapsto F_n(z) \qquad (n = 0, 1, 2, \ldots)$$

suggested by (6.13) and extend it to arbitrary polynomials by the requirement that

(6.15) $$P(w) = \Sigma_{k=0}^n a_k w^k \mapsto (TP)(z) = \Sigma_{k=0}^n a_k F_k(z).$$

This mapping between the polynomials P and TP is one-to-one and onto; for if $TP = 0$, we consider the coefficient of the term of highest degree. Since the degree of F_k is exactly k, we find successively that $a_n = 0$, $a_{n-1} = 0, \ldots$.

If C is rectifiable, both T and its inverse admit an integral representation. Because of (6.8) we have

$$\frac{1}{2\pi i}\int_{|w|=1}\frac{w^n\psi'(w)}{\psi(w)-z}\,dw = F_n(z) \qquad (z \in \operatorname{int} C).$$

and hence

$$(6.16)\ (TP)(z) = \frac{1}{2\pi i}\int_{|w|=1}\frac{P(w)\psi'(w)}{\psi(w)-z}\,dw = \frac{1}{2\pi i}\int_C \frac{P(\phi(\zeta))}{\zeta - z}\,d\zeta \qquad (z \in \operatorname{int} C)$$

for each polynomial P. Furthermore, (6.7) immediately shows that

$$\frac{1}{2\pi i}\int_{|\omega|=1}\frac{(F_n \circ \psi)(\omega)}{\omega - w}\,d\omega = w^n \qquad (w \in \mathbf{D});$$

thus

$$(6.17) \qquad \frac{1}{2\pi i}\int_{|\omega|=1}\frac{(TP)(\psi(\omega))}{\omega - w}\,d\omega = P(w) \qquad (w \in \mathbf{D})$$

for each polynomial P. This gives the desired representation for the inverse mapping.

Now we extend T to a mapping between the Banach spaces $A(\overline{\mathbf{D}})$ and $A(\overline{G})$. As usual, we write

$$A(K) = \{f\colon f \text{ continuous on } K \text{ and analytic in } K^\circ\}$$

and

$$\|f\| = \max\ \{|f(z)|\colon z \in K\}$$

for each compact subset $K \subset \mathbf{C}$. Let Π_n denote the subspace of all polynomials of degree at most n. Then the linear operator T defined by (6.15) is a one-to-one mapping of $\Pi_n \subset A(\overline{\mathbf{D}})$ onto $\Pi_n \subset A(\overline{G})$ ($n = 0, 1, 2, \ldots$). The extension of T mentioned above is based on the following result.

Theorem 2. *If C is a rectifiable Jordan curve of bounded rotation, then the operator T is bounded on $\cup\Pi_n$. For each polynomial P we have*

$$\|TP\| \leqslant (1 + \frac{2V}{\pi})\,\|P\|,$$

where V is the constant defined in (6.11).

Proof. Equation (6.13) implies for $z = \psi(e^{i\theta}) \in C$ that

$$F_n(z) = \frac{1}{\pi}\int_0^{2\pi} e^{int}\, d_t\nu(t,\theta) \qquad (n = 1, 2, \ldots),$$

and therefore

$$\Sigma_{k=0}^{n} a_k F_k(z) = a_0 + \frac{1}{\pi}\int_0^{2\pi} (\Sigma_{k=1}^{n} a_k e^{ikt})\, d_t\nu(t,\theta).$$

This, together with (6.14), implies that

$$\|\Sigma_{k=0}^{n} a_k F_k\| \leqslant |a_0| + (|a_0| + \|\Sigma_{k=0}^{n} a_k w^k\|) \cdot \frac{V}{\pi}.$$

If we take into account that $|a_0| = |P(0)| \leqslant \|P\|$, the assertion follows.

By Theorem 2, we conclude that the domain of T can be extended from $\cup \Pi_n$ to the closure of $\cup \Pi_n$. Since the polynomials are dense in $A(\overline{\mathbf{D}})$, we have obtained an operator T from $A(\overline{\mathbf{D}})$ into $A(\overline{G})$ with the property that

$$(6.18) \qquad \|Tf\| \leqslant (1 + \frac{2V}{\pi})\, \|f\| \quad \text{for each } f \in A(\overline{\mathbf{D}}).$$

Dynkin [1974, p. 270] calls a domain G, on which the operator T is bounded, a *Faber domain*; see also Dynkin [1977].

We observe that by taking limits, we can extend the integral representations (6.16) and (6.17): For each $f \in A(\overline{\mathbf{D}})$ we have

$$(6.16') \qquad (Tf)(z) = \frac{1}{2\pi i}\int_C \frac{f(\phi(\zeta))}{\zeta - z}\, d\zeta \qquad (z \in G).$$

Conversely, for each function Tf, where $f \in A(\overline{\mathbf{D}})$, we have

$$(6.17') \qquad f(w) = \frac{1}{2\pi i}\int_{|\omega|=1} \frac{(Tf)(\psi(\omega))}{\omega - w}\, d\omega \qquad (w \in \mathbf{D}).$$

Using (6.16′) we can show: *If $f \in A(\overline{\mathbf{D}})$ is rational and has a pole of order N at w_0 ($|w_0| > 1$), then Tf is also rational and has a pole of the same order at $\psi(w_0)$.* To see this, we write f as the sum of its principal parts plus a polynomial P. Since TP is a polynomial, it clearly suffices to consider the case $f(w) = 1/(w - w_0)^{n+1}$ for $|w_0| > 1$, $n \geqslant 0$. We then get from (6.16′)

$$(Tf)(z) = \frac{1}{2\pi i}\int_C \frac{f(\phi(\zeta))}{\zeta - z}\, d\zeta = \frac{1}{2\pi i}\int_{C_R} \frac{f(\phi(\zeta))}{\zeta - z}\, d\zeta$$

$$= \frac{1}{2\pi i}\int_{C_R} \frac{f(\omega)\psi'(\omega)}{\psi(\omega) - z}\, d\omega = \frac{1}{2\pi i}\int_{|\omega|=R} \frac{\psi'(\omega)\, d\omega}{(\omega - w_0)^{n+1}[\psi(\omega) - z]},$$

where $1 < R < |w_0|$. If this integral is extended over a circle of radius bigger than $|w_0|$, it will vanish, and therefore the last expression equals

$$-\frac{1}{2\pi i}\int_\gamma \frac{\psi'(\omega)\, d\omega}{[\psi(\omega) - z]\,(\omega - w_0)^{n+1}} = \frac{1}{n!}\frac{\partial^n}{\partial\omega^n}\left[\frac{\psi'(\omega)}{z - \psi(\omega)}\right]_{\omega = w_0},$$

where γ is a small circle around w_0.

This is the representation of $(Tf)(z)$ for $z \in G$, and it proves our assertion. Moreover, it shows that for $f(w) = 1/(w - w_0)^{n+1}$ with $|w_0| > 1$ and $n \geqslant 0$ we have

$$(Tf)(z) = \frac{(\psi'(w_0))^{n+1}}{(z - \psi(w_0))^{n+1}} + \Sigma_{j=-n}^{-1} a_j (z - \psi(w_0))^j.$$

Before returning to approximation theory, we exhibit an additional property of the operator T.

Theorem 3. *If $f \in A(\overline{\mathbf{D}})$ and $f(w) := \Sigma_{n=0}^{\infty} a_n w^n$ ($w \in \mathbf{D}$), then Tf has the Faber coefficients a_n.*

Proof. If we write $f_r(w) := f(rw)$ $(0 < r < 1, w \in \mathbf{D})$, then $f_r \to f$, and hence $Tf_r \to Tf$ $(r \to 1)$. Thus the absolute value of the difference $(Tf)_n - (Tf_r)_n$ of the corresponding Faber coefficients satisfies

$$\left|\frac{1}{2\pi}\int_{|w|=1} [(Tf)(\psi(w)) - (Tf_r)(\psi(w))]\, w^{-n-1} dw\right| \leqslant \|Tf - Tf_r\| \to 0$$

for $r \to 1$. But $(Tf_r)_n = a_n r^n$; for the relation $\Sigma_0^n a_k r^k w^k \Rightarrow f_r(w)$ ($w \in \overline{\mathbf{D}}$, $n \to \infty$) implies that $\Sigma_0^n a_k r^k F_k(z) \Rightarrow (Tf_r)(z)$ ($z \in \overline{G}$, $n \to \infty$); hence $(Tf_r)_n = a_n r^n$ (see the statement following (6.10)). The assertion follows if we let $r \to 1$.

In particular, Theorem 3 implies:

If $Tf = 0$ for some $f \in A(\overline{\mathbf{D}})$, then $f = 0$.

Hence T is a one-to-one mapping of $A(\overline{\mathbf{D}})$ onto the closed subspace $\{F: F = Tf \text{ for } f \in A(\overline{\mathbf{D}})\}$ of $A(\overline{G})$.

D. The quality of approximation inside a curve of bounded rotation

Now we return to our main problem. Throughout Section D we assume that $C \in BR$, $G = \text{int}\, C$, and $F \in A(\overline{G})$. How well can F be approximated on $\overline{G}$ by polynomials of degree n and, if appropriate, which polynomials should be used for the approximation?

D_1. Preparations; uniform convergence

Suppose $F = Tf$ for some $f \in A(\mathbf{D})$. Then we have for arbitrary $a_k^{(n)}$ that

(6.19) $\|F - \Sigma_{k=0}^n a_k^{(n)} F_k\| = \|T(f - \Sigma_{k=0}^n a_k^{(n)} w^k)\| \leqslant \|T\| \cdot \|f - \Sigma_{k=0}^n a_k^{(n)} w^k\|,$

where $\|T\| \leqslant 1 + 2V/\pi$. Our problem is now reduced to the approximation of f in $\overline{\mathbf{D}}$. The following two questions arise:
(i) When does F have the form $F = Tf$ for some $f \in A(\overline{\mathbf{D}})$?
(ii) How is the quality of F reflected in the quality of f?

The next theorem answers question (i).

Theorem 4. *The function $F \in A(\overline{G})$ has the form Tf for some $f \in A(\overline{\mathbf{D}})$ if and only if the Cauchy integral*

$$\Phi(w) := \frac{1}{2\pi i} \int_{|\omega|=1} \frac{h(\omega)}{\omega - w}\, d\omega \quad \textit{with } h = F \circ \psi$$

belongs to $A(\overline{\mathbf{D}})$. If this is the case, then $T\Phi = F$; hence $\Phi = f$.

Proof. If $F = Tf$ for some $f \in A(\overline{\mathbf{D}})$, then Φ must belong to $A(\overline{\mathbf{D}})$ by (6.17′). In this case $\Phi = f$. Conversely, if $\Phi \in A(\overline{\mathbf{D}})$ for a given $F \in A(\overline{G})$, then

$$\Phi(w) = \Sigma_{n=0}^{\infty} w^n \frac{1}{2\pi i} \int_{|\omega|=1} (F \circ \psi)(\omega)\omega^{-n-1}\, d\omega \qquad (w \in \overline{\mathbf{D}}),$$

and Theorem 3 implies that $T\Phi$ and F have the same Faber coefficients. Theorem 1, which was included for this purpose, now yields that $T\Phi = F$; hence F has the required form.

Using Lemma 1, we can reformulate Theorem 4 as follows.

Corollary. *We have $F = Tf$ for some $f \in A(\overline{\mathbf{D}})$ if and only if in addition to $h = F \circ \psi$, its Fourier conjugate $\tilde{h}$ is also continuous on $\partial\mathbf{D}$.*

The next result, concerning the uniform convergence of Faber expansions, follows immediately from (6.19) and Theorem 4.

Theorem 5 (Kövari and Pommerenke [1967]). *The Faber expansion of $F \in A(\overline{G})$ converges uniformly in $\overline{G}$ if the corresponding Cauchy integral Φ has a power series expansion that converges uniformly in $\overline{\mathbf{D}}$. This will be the case if both h and $\tilde{h}$ have uniformly convergent Fourier series.*

D_2. The modulus of continuity of the Cauchy integral corresponding to *h*

In order to obtain statements about the quality of approximation, one must assume more about h. First let $\omega_p(h, t)$ denote the pth*-order modulus of continuity of the function* h $(p = 1, 2, \ldots)$; see Timan [1963, p. 102]. For example,

$$\omega_1(h, t) = \sup\left\{|h(e^{i\phi}) - h(e^{i(\phi+\delta)})|: |\delta| \leqslant t,\ |\phi| \leqslant \pi\right\}$$

and

$$\omega_2(h, t) = \sup\left\{|h(e^{i\phi}) - 2h(e^{i(\phi+\delta)}) + h(e^{i(\phi+2\delta)})|: |\delta| \leqslant t,\ |\phi| \leqslant \pi\right\}.$$

In the following it is important that the modulus of continuity of $\tilde{h}$ can be estimated by that of h (Timan [1963, p. 162]). For if $\int_0^1 \omega_p(h, u)u^{-1}du < \infty$, then $\tilde{h}$ is continuous, and there exists an absolute constant c_p such that

$$\text{(6.20)} \quad \omega_p(\tilde{h}, t) \leqslant c_p\left[\int_0^t \frac{\omega_p(h, u)}{u}\,du + t^p \int_t^1 \frac{\omega_p(h, u)}{u^{p+1}}\,du\right] \quad (p = 1, 2, \ldots).$$

This implies the following result.

Theorem 6. *If the function $h = F \cdot \psi$ has a modulus of continuity $\omega_p(h, t)$ on $\partial\mathbf{D}$ such that*

$$\int_0^1 \omega_p(h, u)u^{-1}du < \infty,$$

then the Cauchy integral Φ corresponding to h is continuous in $\overline{\mathbf{D}}$ and has the modulus of continuity

$$\text{(6.21)} \quad \omega_p(\Phi, t) \leqslant C_p\left[\omega_p(h, t) + \int_0^t \frac{\omega_p(h, u)}{u}\,du + t^p \int_t^1 \frac{\omega_p(h, u)}{u^{p+1}}\,du\right]$$

on $\partial\mathbf{D}$. Here the C_p are certain constants independent of h.

Proof. Because $\int_0^1 \omega_p(h, u)u^{-1}du < \infty$, the continuity of h on $\partial\mathbf{D}$ implies that of $\tilde{h}$. Representation (6.3) for $\Phi(e^{i\phi})$, together with (6.20), establishes the assertion.

We call attention to two *special cases*. a) If $h \in \text{Lip}\,\alpha$ $(0 < \alpha < 1)$, that is, if $\omega_1(h, t) \leqslant \text{const} \cdot t^{\alpha}$, then also $\Phi \in \text{Lip}\,\alpha$. See also a corresponding theorem for general Cauchy integrals in the book by Privalov [1956, p. 143]. b) If $h \in Z$ (the Zygmund class), that is, if $\omega_2(h, t) \leqslant \text{const} \cdot t$, then also $\Phi \in Z$.

D_3. The quality of the approximation

For functions $\Phi \in A(\overline{\mathbf{D}})$, approximation results involving the modulus of continuity are well known (theorems of "Jackson type"). There exist absolute constants C_p such that

$$\|\Phi - P_n\| \leqslant C_p\, \omega_p(\Phi, 1/n) \qquad (n = 1, 2, \ldots)$$

for certain polynomials P_n of degree n. Their images $TP_n \in \Pi_n$ yield a corresponding approximation of $T\Phi = F$.

These P_n can be obtained explicitly from the power series expansion $\Phi(w) = \Sigma_{n=0}^{\infty} a_n w^n$ $(w \in \mathbf{D})$ by various methods. Here we point out the following procedure. To a series $\Sigma_{j=0}^{\infty} a_j$ we assign (for fixed $p \in \mathbf{N}$) the transformations

$$\tau_n^{(p)} := \frac{1}{(n+1)^p}\, \Sigma_{j=0}^{n}\, [(n+1)^p - j^p]\, a_j \qquad (n = 0, 1, \ldots).$$

For $p = 1$, these are the Fejér means σ_n of Σa_j. The application of these transformations to $\Phi(w) = \Sigma_{j=0}^{\infty} a_j w^j$ thus yields the nth degree polynomials

$$\tau_n^{(p)}(w) = \frac{1}{(n+1)^p}\, \Sigma_{j=0}^{n}\, [(n+1)^p - j^p]\, a_j w^j.$$

About these it is known (Gaier [1977, p. 4]) that

$$\|\Phi - \tau_n^{(p)}\| \leqslant C_p \omega_p(\Phi, 1/n) \qquad (n = 1, 2, \ldots).$$

Their images under the mapping T have the form

$$(6.22)\quad T_n^{(p)}(z) := \frac{1}{(n+1)^p}\, \Sigma_{j=0}^{n}\, [(n+1)^p - j^p]\, a_j F_j(z) \qquad (n = 0, 1, 2, \ldots),$$

and our results can now be summarized as follows.

Theorem 7. *Suppose C is a rectifiable Jordan curve of bounded rotation V. Suppose F is analytic in G = int C, continuous in $\bar{G}$, and suppose that for $p \in \mathbf{N}$ the function $h = F \cdot \psi$ has the modulus of continuity $\omega_p(h, t)$, where $\int_0^1 \omega_p(h, t)\, t^{-1} dt < \infty$. If we now use the Faber coefficients a_j of F to construct the polynomials $T_n^{(p)}$ of degree n according to* (6.22), *then there exist absolute constants D_p such that*

$$\|F - T_n^{(p)}\| \leqslant D_p \cdot V \cdot \omega_p(\Phi, 1/n) \qquad (n = 1, 2, \ldots).$$

Here $\omega_p(\Phi, t)$ can be estimated according to (6.21).

We call attention to two *special cases.* a) $h \in \mathrm{Lip}\,\alpha$ $(0 < \alpha < 1)$: Then (see above) we also have $\Phi \in \mathrm{Lip}\,\alpha$, that is, $\omega_1(\Phi, 1/n) \leqslant \text{const} \cdot n^{-\alpha}$. In this case, the Fejér means $T_n^{(1)}$ of the Faber expansion of F satisfy

$$\|F - T_n^{(1)}\| \leqslant \text{const} \cdot n^{-\alpha} \qquad (n = 1, 2, \ldots).$$

b) $h \in Z$: Then (see above) we also have $\Phi \in Z$, that is, $\omega_2(\Phi, 1/n) \leqslant \text{const} \cdot n^{-1}$. In this case, the means $T_n^{(2)}$ of the Faber expansion of F satisfy

$$\|F - T_n^{(2)}\| \leqslant \text{const} \cdot n^{-1} \qquad (n = 1, 2, \ldots).$$

In particular, since $\mathrm{Lip}\,1 \subset Z$, the last inequality is satisfied if $h \in \mathrm{Lip}\,1$.

The special case where $h \in \mathrm{Lip}\,\alpha$ $(0 < \alpha \leqslant 1)$ has been treated by Ganelius [1973] in a different (nonconstructive) way. For his results, Kövari [1972] uses the de la Vallée-Poussin means, which also appear in Švai's work [1973]; however, Kövari's results are more general. Fejér means also have been used earlier: Sewell [1942], Al'per [1955], and Dincen [1964]. In these works the Jordan curve C is always assumed to be smooth. Bruĭ [1976] allows corners.

If one wishes to estimate $\|F - T_n^{(p)}\|$ directly by means of the modulus of continuity of F (rather than that of h), one must make further assumptions about C. We say that

$C \in K_1$ if C is a convex Jordan curve,

$C \in K_\alpha$ $(0 < \alpha < 1)$ if C is a piecewise convex Jordan curve whose smallest exterior angle is $\pi\alpha$.

If $C \in K_\alpha$, then $\psi \in \mathrm{Lip}\,\alpha$ on $\partial\mathbf{D}$ $(0 < \alpha \leqslant 1)$; see Kövari [1972]. Theorem 7 now implies the following.

Theorem 8. *Suppose that* $C \in K_\alpha$ $(0 < \alpha \leqslant 1)$, *F is analytic in* $G = \text{int}\, C$ *and continuous in* $\overline{G}$, *and*

$$|F(z_1) - F(z_2)| \leqslant \text{const} \cdot |z_1 - z_2|^\beta \quad \textit{for } z_1, z_2 \in C, 0 < \beta \leqslant 1.$$

Then the polynomials of degree n defined by (6.22) *satisfy*

$$\|F - T_n^{(1)}\| \leqslant O(1) \cdot n^{-\alpha\beta} \quad \textit{if } \alpha\beta < 1$$

and

$$\|F - T_n^{(2)}\| \leqslant O(1) \cdot n^{-1} \quad \textit{if } \alpha\beta = 1.$$

Proof. The assumptions imply that $h = F \circ \psi \in \text{Lip}\, \alpha\beta$. The special cases mentioned after Theorem 7 now establish the assertion. – Recall that the $T_n^{(1)}$ are the Fejér means of the Faber expansion of F.

E. Report on additional results

E_1. Additional uniform estimates

Here we refer primarily to a paper by Lesley, Vinge, and Warschawski [1974]. The curve C is required to be rectifiable and to satisfy a "c-condition": If z_1, z_2 are two arbitrary points on C and $\Delta(z_1, z_2)$ denotes the length of the shorter arc from z_1 to z_2, then

$$\Delta(z_1, z_2) \leqslant c|z_1 - z_2| \quad \text{for some fixed } c > 1.$$

With this assumption, it is shown that for the partial sums S_n of the Faber expansion of F the inequalities

$$\|F - S_n\| \leqslant [A(\log n)^2 + B]\, E_n(F) \qquad (n = 1, 2, \ldots)$$

hold, where A and B are constants and $E_n(F) = \inf\{\|F - P\| : P \in \Pi_n\}$ denotes the minimal error in approximating F by polynomials of degree at most n. (Kövari and Pommerenke [1967] have only $\log n$ on the right-hand side if $C \in BR$.) In addition $\|F - J_n\|$ is estimated, where the J_n are the Jackson sums of the Faber expansion of F. By the way, for each Jordan curve we have $\|F - S_n\| \leqslant An^\alpha E_n(F)$, where A and $\alpha \in (0.138, 0.5)$ are universal constants; see Kövari and Pommerenke [1967, p. 198]. This is obtained by estimating the "Lebesgue constants"

$$L_n := \frac{1}{2\pi} \int_{|\omega|=1} |\Sigma_{k=0}^{n} \frac{F_k(\psi(\omega))}{\omega^{k+1}}| \, |d\omega| \qquad (n = 1, 2, \ldots).$$

E_2. Local estimates

The previous estimates use only the global modulus of continuity of $F \circ \psi$ to judge the quality of the approximation. But with the modulus of continuity of F given, the approximation will be better on smooth parts of C. More detailed investigations about this have been carried out by a great many Soviet mathematicians; see the remarks following this section. Here we shall emphasize a new, especially general result by Belyĭ [1977].

Suppose C is some arbitrary Jordan curve and C_R is the level curve corresponding to $R > 1$, that is,

$$C_R = \{z: |\phi(z)| = R > 1\},$$

where ϕ is again the conformal mapping, normalized at ∞, of ext C onto $\{w: |w| > 1\}$. Write

$$d_R(z) = \text{dist}\,(z, C_R) \quad \text{for } z \in C.$$

Furthermore, C is called a *quasiconformal curve* if there exists a quasiconformal homeomorphism of $\mathbb{C}$ onto $\mathbb{C}$ that maps $\partial\mathbf{D}$ onto C. Such curves can be characterized geometrically: For two arbitrary points z_1, z_2 on C we consider the arc from z_1 to z_2 with the smaller diameter; suppose this diameter $d(z_1, z_2)$ satisfies $d(z_1, z_2) \leqslant k|z_1 - z_2|$ for some constant k. See, for example, Lehto and Virtanen [1965, p. 101 ff.]. The curve C need not be rectifiable. With this notation, Belyĭ [1977, p. 333] proved the following.

Suppose C is a quasiconformal curve, $G = \text{int}\, C$, and $F \in A(\overline{G})$. Let $\omega(F, t)$ denote the modulus of continuity of F on $\overline{G}$. Then there exist polynomials P_n of degree n for which

$$|F(z) - P_n(z)| \leqslant M\omega(F, d_{1+1/n}(z)) \qquad (z \in C), \tag{6.23}$$

where the constant M is independent of z and n.

The estimate is better, the closer the level curves $C_{1+1/n}$ are to a given point $z_0 \in C$. If, say, C is a polygon and z_0 is a corner with exterior angle $\alpha\pi$, then

$$d_R(z_0) \sim (R - 1)^{\alpha},$$

and thus

$$|F(z_0) - P_n(z_0)| = O(1)\omega(F, c/n^{\alpha}) \qquad (n \to \infty).$$

Belyĭ also proves inverse theorems. Other authors consider $|F^{(r)}(z) - P_n^{(r)}(z)|$ provided that $F^{(r)} \in A(\bar{G})$. As a rule, such proofs are technically extremely complicated.

Remarks about §6

1. We used the boundedness of the operator T if $C \in BR$. This was already the basis of Kövari's work [1972], and it is pursued further by Andersson [1975], [1976].

2. As was mentioned before, many Soviet authors are concerned with uniform estimates or estimates of the form (6.23). In addition to the older survey article by Mergelyan [1952, Chapter III] and the already cited literature, we refer to the following newer articles:

Andersson [1975], [1976]; Andraško [1964]; Belyĭ [1965]; Belyĭ and Mikljukov [1974]; Dzjadyk [1962], [1963a], [1963b], [1966], [1968], [1969], [1972], [1975], [1977]; Dzjadyk and Alibekov [1968]; Dzjadyk and Galan [1965]; Dzjadyk and Švai [1971]; Kolesnik and Andraško [1971]; Lebedev and Širokov [1971]; Lebedev and Tamrazov [1970]; Ševčuk [1973]; Širokov [1971], [1972], [1974a], [1974b], [1976]; Tamrazov [1971], [1973], [1975].

Bijvoets, Hogeveen, and Korevaar [1980] give simpler proofs of the inverse theorems of Lebedev and Tamrazov in the case where the complement of the compact set K is simply connected. The recently published book by Dzjadyk [1977, Chapter 9, §6–10] presents his earlier results in detail.

3. In the case where $C \in K_1$, we used in the proof of Theorem 8 that $\psi \in$ Lip 1 on $\partial\mathbb{D}$. This follows from the boundedness of $\psi'(w)$ for $|w| > 1$. And the latter property can be proved in different ways: If, say, the curve C has capacity 1, then Löwner [1919, p. 76] already proved the sharp estimate $|\psi'(w)| \leqslant 1 + |w|^{-2}$ $(|w| > 1)$. Kövari and Pommerenke [1967, p. 195] show in another way that $|\psi'(w) - 1| \leqslant |w|^{-2}$ $(|w| > 1)$; the integral representation of $\log \psi'$ also yields immediately that ψ' is bounded (Kövari [1972, p. 370]). Finally, if C is convex, the function $w\psi'(w)$ is schlicht for $|w| > 1$ and normalized at ∞ such that $w\psi'(w) = w + a_1/w + \ldots$. Hence a distortion-type theorem ("Verschiebungssatz" – Grötzsch; Goluzin [1969, p. 136]) can be applied, and we find $|w\psi'(w) - w| \leqslant |w|^{-1}$, that is, $|\psi'(w) - 1| \leqslant |w|^{-2}$ $(|w| > 1)$.

If C satisfies a c-condition (see Section E_1), then $\psi \in$ Lip α, where

$\alpha = \dfrac{2}{(1+c)^2}$; use Lemma 1 in Warschawski [1961, p. 615]. And if C is a K-quasiconformal curve (the image of $\partial\mathbb{D}$ under a K-quasiconformal mapping of $\mathbb{C}$ onto $\mathbb{C}$), then $\psi \in$ Lip α, where $\alpha = 1/K^2$. This follows from theorems about quasiconformal mappings.

Chapter II

APPROXIMATION BY INTERPOLATION

Next to series expansion, interpolation represents a further important tool for the approximation of functions; we now turn to this method.

As far as literature is concerned, we cite in particular Davis [1963, Chapter IV], Smirnov and Lebedev [1968, Chapter I], and the book by Walsh [1969].

§1. Hermite's interpolation formula

A. The interpolating polynomial

If $n + 1$ pairs (z_k, w_k) $(k = 0, 1, \ldots, n)$ of complex numbers are given, where the z_k are to be distinct for now, then there exists exactly one polynomial P of degree (at most) n such that

$$P(z_k) = w_k \qquad (k = 0, 1, \ldots, n). \tag{1.1}$$

One way of obtaining this polynomial is through *Lagrange's formula of interpolation*. To this end, we let

$$\omega(z) = \Pi_{k=0}^{n} (z - z_k) \ \text{ and } \ l_k(z) = \frac{\omega(z)}{\omega'(z_k)(z - z_k)} \qquad (k = 0, 1, \ldots, n).$$

Each of these basic polynomials l_k is exactly of degree n, and we have

$$l_k(z_j) = \begin{cases} 1 & \text{if } j = k, \\ 0 & \text{if } j \neq k. \end{cases}$$

Hence the nth degree polynomial

$$L_n(z) = \Sigma_{k=0}^{n} w_k l_k(z) \tag{1.2}$$

satisfies the interpolation requirement (1.1).

The case where $w_k = f(z_k)$, with a function f analytic in a domain G and with interpolation points $z_k \in G$, is important for our purposes. Here the interpolating polynomial can also be represented by a complex integral. Suppose the boundary ∂G of G consists of finitely many rectifiable Jordan curves with positive orientation relative to G, and suppose f is analytic in G and continuous in $\bar{G}$. The interpolation problem is: $P(z_k) = f(z_k)$, where $z_k \in G$ $(k = 0, 1, \ldots, n)$. It is solved by the formula

$$L_n(z) = \frac{1}{2\pi i} \int_{\partial G} \frac{\omega(t) - \omega(z)}{t - z} \cdot \frac{f(t)}{\omega(t)}\, dt \qquad (z \in G), \tag{1.3}$$

and we have that

$$f(z) - L_n(z) = \frac{1}{2\pi i} \int_{\partial G} \frac{\omega(z)}{\omega(t)} \cdot \frac{f(t)}{t - z}\, dt \qquad (z \in G). \tag{1.4}$$

To begin with, it is clear that (1.3) represents a polynomial of degree n; (1.4) follows from (1.3) because $f(z) = \frac{1}{2\pi i} \int_{\partial G} \frac{f(t)}{t - z}\, dt$; and (1.4) implies that $f(z) - L_n(z) = \omega(z) \cdot h(z)$, where the function h is analytic in G. Hence $f - L_n$ vanishes at the points z_k.

Relation (1.3) is the *Hermitian representation* of the interpolating polynomial, and (1.4) is an integral representation of the interpolation error. If several, say m, of the z_k coincide, we have "m-fold interpolation." This means that $f - L_n$ has a zero of order m at the corresponding point. Formulas (1.3) and (1.4) are also valid in this case. In addition, we have the following result.

Corollary. *Formulas* (1.3) *and* (1.4) *remain valid if* $G = G_1 \cup G_2 \cup \ldots \cup G_N$, *where the domains* G_j *are disjoint and of the form described above.*

This corollary is important when we wish to approximate simultaneously by a single interpolating polynomial analytic functions defined in several domains.

B. Special cases of Hermite's formula

We now deal with three special cases of (1.3). Suppose first that f is analytic in $\mathbf{D} = \{z: |z| < 1\}$ and that $z_k = 0$ $(k = 0, 1, \ldots, n)$. Then $\omega(z) = z^{n+1}$, and the interpolating polynomial is

$$L_n^{(1)}(z) = \frac{1}{2\pi i}\int_{|t|=r} \frac{t^{n+1} - z^{n+1}}{t - z} \cdot \frac{f(t)}{t^{n+1}}\, dt$$

$$= \frac{1}{2\pi i}\int_{|t|=r} \frac{1 - (z/t)^{n+1}}{1 - (z/t)} \cdot \frac{f(t)}{t}\, dt \qquad (0 < r < 1).$$

The first quotient in the second integral equals $1 + (z/t) + \ldots + (z/t)^n$, and hence the polynomial $L_n^{(1)}(z) = \Sigma_{j=0}^n a_j z^j$ is the nth partial sum of the power series of f at 0. (This follows also from the fact that $f(z) - \Sigma_{j=0}^n a_j z^j$ has a zero of order $n + 1$ at $z = 0$.)

Second, suppose that the $z_k (k = 0, 1, \ldots, n)$ are the $(n + 1)^{\text{st}}$ roots of unity and that f is analytic in $G = \{z: |z| < R\}$ and continuous in $\overline{G}$ for some $R > 1$. Now

$$\omega(z) = \Pi_{k=0}^n (z - z_k) = z^{n+1} - 1,$$

and the interpolating polynomial has the form

$$L_n^{(2)}(z) = \frac{1}{2\pi i}\int_{|t|=R} \frac{t^{n+1} - z^{n+1}}{t - z} \cdot \frac{f(t)}{t^{n+1} - 1}\, dt \qquad (|z| < R).$$

This expression is similar to $L_n^{(1)}$; therefore we consider

$$L_n^{(2)}(z) - L_n^{(1)}(z) = \frac{1}{2\pi i}\int_{|t|=R} \frac{t^{n+1} - z^{n+1}}{t^{n+1}(t^{n+1} - 1)(t - z)} f(t)dt.$$

This representation is certainly valid for $|z| < R$; but since both sides are polynomials in z that agree for $|z| < R$, the representation is actually valid in $\mathbb{C}$. For $|z| = \rho > R$, the right-hand side obviously is

$$O(1) \cdot (\rho^n / R^{2n}) \qquad (n \to \infty),$$

and this tends to zero as long as $\rho < R^2$. We thus have the following theorem.

Theorem 1 (Walsh [1969, p. 153]). *The interpolating polynomials $L_n^{(1)}$ and $L_n^{(2)}$ are equiconvergent on* $\{z: |z| \leqslant \rho\}$ $(R < \rho < R^2)$:

$$(1.5) \qquad L_n^{(2)}(z) - L_n^{(1)}(z) \Rightarrow 0 \qquad (n \to \infty;\ |z| \leqslant \rho, R < \rho < R^2).$$

For the assertion in $\{z: |z| < \rho\}$ we have used the maximum principle. For example, if f is analytic in $\bar{G}$, then $L_n^{(1)}(z) \Rightarrow f(z)$ implies that also $L_n^{(2)}(z) \Rightarrow f(z)$ $(z \in \bar{G}, n \to \infty)$.

Finally, our third case deals with interpolation on the interval $[-1, +1]$. As interpolation points $z_0, z_1, \ldots, z_{n-1}$ we use the n zeros of the nth Chebyshev polynomial

$$\omega(x) = \Pi_{k=0}^{n-1}(x - z_k) = \frac{1}{2^{n-1}} \cos(n \arccos x) =: \tilde{T}_n(x).$$

For our purposes it is now important to know the behavior of $\omega(z)$ for complex values of z. To this end, we verify first that

$$\omega(z) = \frac{1}{2^n}\left(w^n + \frac{1}{w^n}\right), \quad \text{where } z = \frac{1}{2}\left(w + \frac{1}{w}\right). \tag{1.6}$$

For if one substitutes $z = \frac{1}{2}(w + w^{-1})$ in the polynomial $\omega(z)$, one obtains a function of w that is analytic in $\mathbb{C}\setminus\{0\}$; the same is true for the right-hand side of (1.6). For $w = e^{i\phi}$, both sides equal $(\cos n\phi)/2^{n-1}$, and the identity theorem implies that (1.6) holds in general.

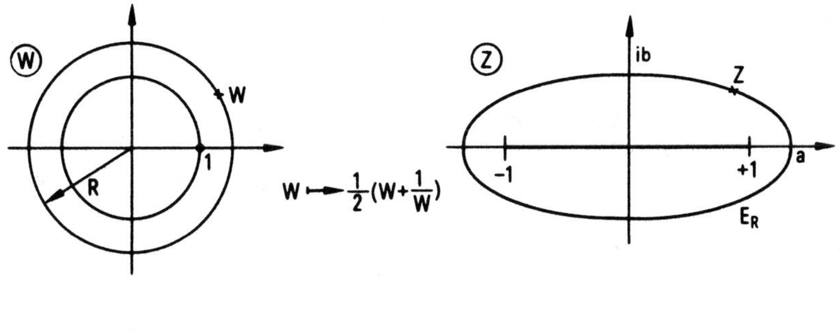

The image of $\{w: |w| = R\}$ $(R > 1)$ is the ellipse E_R with semiaxes $a = \frac{1}{2}(R + R^{-1})$ and $b = \frac{1}{2}(R - R^{-1})$. Further, (1.6) implies that

$$|\omega(z)| = \frac{R^n}{2^n} \, |1 + w^{-2n}| \quad \text{if } z \in E_R. \tag{1.7}$$

Now let G_R denote the interior of the ellipse E_R, and suppose f is analytic in $\bar{G}_R$. Let $L_{n-1}^{(3)}$ denote the interpolating polynomial of degree $n - 1$ corresponding to the interpolation points $z_0, z_1, \ldots, z_{n-1}$. For $-1 \leqslant x \leqslant +1$, relations (1.4) and (1.7) immediately imply the estimate

$$f(x) - L_{n-1}^{(3)}(x) = \frac{1}{2\pi i} \int_{E_R} \frac{\omega(x)}{\omega(t)} \cdot \frac{f(t)}{t-x}\, dt$$

$$= O(1) \cdot \frac{1}{2^n} \cdot \frac{2^n}{R^n} = O\left(\frac{1}{R^n}\right) \qquad (n \to \infty).$$

The larger the domain surrounding $[-1, +1]$ is in which f is analytic, the faster the convergence $L_n^{(3)} \Rightarrow f$ takes place on $[-1, +1]$.

§2. Interpolation in uniformly distributed points; Fejér points, Fekete points

We now begin to discuss the following problem: Under which assumptions about f and about the location of the interpolation points can one prove that $L_n(z) \to f(z)$ $(n \to \infty)$? What can be said about the rate of convergence?

A. Preparations; rough statement about convergence

Throughout §2, with the exception of the beginning of Part D, we assume K is a compact subset of $\mathbb{C}$ whose complement $K^c = \mathbb{C} \backslash K$ is a simply connected domain. Then there exists a conformal mapping

$$z = \psi(w) = cw + c_0 + c_1/w + \ldots \qquad (c > 0),$$

normalized at ∞, of $\{w\colon |w| > 1\}$ onto K^c; here c is called the capacity of ∂K. Let ϕ denote its inverse function. Further, let

$$C_\rho = \{z\colon |\phi(z)| = \rho\} \qquad (\rho > 1)$$

denote the level curves in K^c.

Since we wish to interpolate a function f defined on K, we suppose that for each n $(n = 0, 1, \ldots)$, there are $n + 1$ given points of interpolation $z_k^{(n)} \in K$ $(k = 0, 1, \ldots, n)$. All together, we have a "node matrix":

$$\begin{array}{llll} z_0^{(0)} & & & \\ z_0^{(1)} & z_1^{(1)} & & \\ \cdots\cdots & & & \\ z_0^{(n)} & z_1^{(n)} & \cdots & z_n^{(n)} \\ \cdots\cdots\cdots & & & \end{array}$$

of points in K, and we write

$$\omega_n(z) := \Pi_{k=0}^{n} (z - z_k^{(n)}) \qquad (n = 0, 1, 2, \ldots).$$

If f is analytic on K, hence also inside and on a suitable rectifiable Jordan curve C with $K \subset \operatorname{int} C$, then, according to (1.4), the interpolation error is

$$f(z) - L_n(z) = \frac{1}{2\pi i} \int_C \frac{\omega_n(z)}{\omega_n(t)} \cdot \frac{f(t)}{t - z}\, dt \qquad (z \in K).$$

This can be used to obtain a rough statement about the convergence. For if

(2.1) $$\operatorname{diam} K =: D_1 < D_2 := \operatorname{dist}(K, C),$$

then $|\omega_n(z)| \leqslant D_1^{n+1}$ $(z \in K)$ and $|\omega_n(t)| \geqslant D_2^{n+1}$ $(t \in C)$. Therefore we have, in this case,

$$L_n(z) \Rightarrow f(z) \qquad (n \to \infty; z \in K),$$

and the convergence is *independent* of the choice of the nodes $z_k^{(n)} \in K$.

Example. If f is analytic inside and on C (see the following figure), then we have for an arbitrary choice of the nodes on $[-1, +1]$ that

$$L_n(z) \Rightarrow f(z) \qquad (n \to \infty; z \in [-1, 1]).$$

Here the constant "2" is best possible, as one can verify easily.

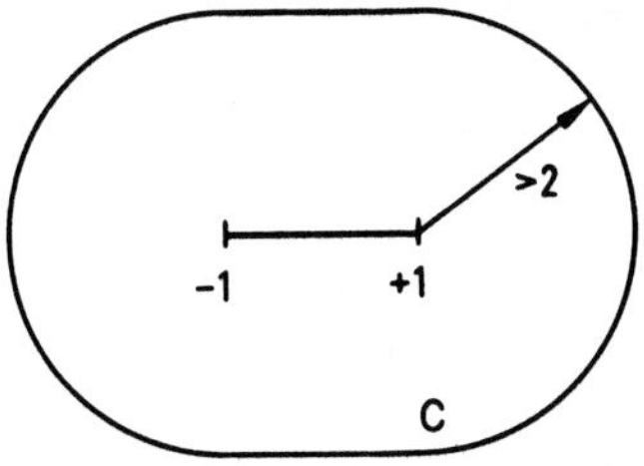

If the region around K in which f is still analytic is smaller than indicated by (2.1), then the nodes $z_k^{(n)}$ in K must be distributed more carefully. This will be done in the following.

B. General convergence theorem of Kalmár and Walsh

First we define what will be meant by a uniform distribution. To this end we consider the numbers

$$M_n = \max \{|\omega_n(z)|: z \in K\} \quad (n = 0, 1, 2, \ldots);$$

the maximum is attained on $\partial K = \partial K^c$. *The inequality*

$$M_n \geqslant c^{n+1} \quad (n = 0, 1, 2, \ldots) \tag{2.2}$$

always holds.

To prove the last statement, we consider the auxiliary function

$$H_n(z) = \frac{\omega_n(z)}{[c\phi(z)]^{n+1}} \quad \text{for } z \in K^c.$$

The function $H_n(z)$ is analytic in K^c and tends to 1 as $z \to \infty$. According to the maximum principle, we have that

$$\max \{|H_n(z)|: z \in C_\rho\} \geqslant 1 \quad \text{for each } \rho > 1;$$

hence

$$\max \{|\omega_n(z)|: z \in C_\rho\} \geqslant (c\rho)^{n+1} > c^{n+1}.$$

Since M_n is attained on ∂K^c, assertion (2.2) follows if we let $\rho \to 1$.

This leads to the following definition.

Definition 1. *The nodes $z_k^{(n)}$ are uniformly distributed on K if*

$$\sqrt[n+1]{M_n} \to c \quad (n \to \infty). \tag{2.3}$$

Example. We take $K = [-1, +1]$, and as nodes on K we choose the zeros of the $(n + 1)^{\text{st}}$ Chebyshev polynomial. Then

$$\omega_n(x) = 2^{-n} \cos[(n + 1) \arccos x], \quad M_n = 2^{-n},$$

and now condition (2.3) for uniform distribution is satisfied, because $\psi(w) = \tfrac{1}{2}(w + w^{-1})$ and therefore $c = \tfrac{1}{2}$.

Condition (2.3) can be expressed differently with the help of the functions

$$\theta_n(z) := \sqrt[n+1]{H_n(z)} = \frac{\sqrt[n+1]{\omega_n(z)}}{c\phi(z)}. \tag{2.4}$$

These functions are analytic in K^c, since the zeros of ω_n are in K. Further, $\lim_{z\to\infty}\theta_n(z) = 1$ for each n, so that the functions θ_n are uniformly bounded on each compact subset of K^c and hence form a normal family there.

Lemma 1. *Relation* (2.3) *holds if and only if* $\theta_n(z) \rightrightarrows 1$ $(n \to \infty)$ *on each compact subset of* K^c.

Proof. a) If $|\theta_n(z)| \rightrightarrows 1$ $(n \to \infty; z \in C_\rho)$ for each $\rho > 1$, then

$$\sqrt[n+1]{\max\{|\omega_n(z)|: z \in C_\rho\}} \to c\rho \qquad (n \to \infty),$$

and the left-hand side exceeds $\sqrt[n+1]{M_n}$. It follows that $\limsup \sqrt[n+1]{M_n} \leqslant c\rho$ for each $\rho > 1$, and considering (2.2) we obtain (2.3).

b) For each n, the maximum principle yields

$$|\theta_n(z)| \leqslant \sqrt[n+1]{M_n}/c \quad \text{for} \quad z \in K^c.$$

If the right-hand side tends to 1, then each limit function θ of the normal family $\{\theta_n\}$ satisfies the condition $|\theta(z)| \leqslant 1$ for $z \in K^c$, and also $\lim_{z\to\infty}\theta(z) = 1$. It follows that $\theta(z) = 1$ for $z \in K^c$; hence $\theta_n \rightrightarrows 1$ on each compact subset of K^c.

Now we turn to the connection between the uniform distribution of the nodes and the convergence of the corresponding interpolation process.

Theorem 1 (Kalmár 1926, Walsh 1933). *The convergence* $L_n(z) \rightrightarrows f(z)$ $(n \to \infty; z \in K)$ *takes place for each function f analytic on K if and only if the interpolation nodes* $z_k^{(n)}$ *are uniformly distributed on K.*

Later (Theorem 2), we shall make an additional statement about the rate of convergence.

Proof. a) The uniform distribution of the nodes is *necessary*. i) For each $z_0 \in K^c$ the function $f_0(z) = 1/(z_0 - z)$ is analytic on K; hence, by assumption

$$L_n(z, f_0) \rightrightarrows f_0(z) \qquad (n \to \infty; z \in K).$$

Here L_n has the representation

$$L_n(z, f_0) = \frac{1}{z_0 - z} - \frac{\omega_n(z)}{\omega_n(z_0)(z_0 - z)},$$

as one can verify immediately. This implies that

$$M_n/|\omega_n(z_0)| \to 0 \qquad (n \to \infty)$$

for each fixed $z_0 \in K^c$.

ii) Now suppose $\limsup \sqrt[n+1]{M_n} > c$; that is, there exists an $\epsilon > 0$ such that

$$\sqrt[n+1]{M_n} \geqslant (1+\epsilon)c \quad \text{if } n = n_k \to \infty.$$

In this case we would choose $\rho \in (1, 1+\epsilon)$, select an arbitrary z_0 on C_ρ, and obtain – see (2.4) –

$$|\theta_n(z_0)| = \frac{\sqrt[n+1]{|\omega_n(z_0)|}}{c|\phi(z_0)|} \geqslant \frac{\sqrt[n+1]{M_n}}{c\rho} \geqslant \frac{(1+\epsilon)c}{c\rho} = \frac{1+\epsilon}{\rho}$$

for $n = n_k$ and such indices k for which $M_{n_k} \leqslant |\omega_{n_k}(z_0)|$. A convergent subsequence of $\{\theta_{n_k}\}$ therefore has a limit function θ with the property that

$$|\theta(z)| \geqslant \frac{1+\epsilon}{\rho} > 1 \quad \text{for each } z \in C_\rho,$$

contradicting $\theta(\infty) = 1$ (minimum principle!). Note that the functions θ_n have no zeros in K^c.

b) The uniform distribution of the nodes is *sufficient*. We actually prove an even stronger statement.

Theorem 2. *Suppose $\rho > 1$ is the largest number such that f is analytic inside C_p. The interpolating polynomials L_n with nodes $z_k^{(n)}$ that are uniformly distributed on K then satisfy the condition*

$$\overline{\lim} \sqrt[n]{\max\{|f(z) - L_n(z)| : z \in K\}} = 1/\rho; \tag{2.5}$$

that is, the sequence $\{L_n\}$ converges maximally on K to f (see Chapter I, §4, C).

For each number $R < \rho$ we thus have

$$\max_{z \in K} |f(z) - L_n(z)| = O(R^{-n}) \qquad (n \to \infty).$$

This fact was needed in Chapter I, §4 for the proof of Theorem 4. However, we still need to prove the existence of points that are uniformly distributed on K.

Proof of Theorem 2. For each $R \in (1, \rho)$, Lemma 1 implies that

$$\sqrt[n+1]{|\omega_n(t)|} \Rightarrow cR \qquad (t \in C_R; n \to \infty).$$

In addition, we have

$$|\omega_n(z)| \leqslant M_n \qquad (z \in K; n = 0, 1, \ldots).$$

Hermite's formula (1.4) thus allows us to conclude that

$$f(z) - L_n(z) = \frac{1}{2\pi i} \int_{C_R} \frac{\omega_n(z)}{\omega_n(t)} \cdot \frac{f(t)}{t - z}\, dt$$

$$= O(1) \cdot \frac{M_n}{(cR - \epsilon)^n} = O(1) \cdot \left(\frac{c + \epsilon}{cR - \epsilon}\right)^n \qquad (z \in K; n \to \infty)$$

for each $\epsilon > 0$; here we have used (2.3). From this, (2.5) follows immediately with "$\leqslant$" instead of "=". We have already seen in Chapter 1, §4, C that strict inequality is impossible. Otherwise f would have an analytic continuation beyond C_ρ. This establishes Theorem 2, and hence also Theorem 1.

We now turn to the problem of obtaining systems of uniformly distributed points.

C. The system of Fejér points

Concerning the compact set K, we now require that the mapping ψ from $\{w\colon |w| > 1\}$ onto K^c has a *continuous* extension to $\{w\colon |w| \geqslant 1\}$; this will be the case, for example, if ∂K is a Jordan curve or a Jordan arc. Then the points

$$z_k^{(n)} = \psi(e^{2\pi i k/(n+1)}) \qquad (k = 0, 1, \ldots, n)$$

are called the *Fejér points of order n* on K. They are the images under ψ of the $(n + 1)^{\text{st}}$ roots of unity.

Theorem 3 (Fejér 1918). *The Fejér points are uniformly distributed on K.*

Proof. In view of (2.2) we need to prove only that

$$\limsup \sqrt[n+1]{M_n} \leqslant c$$

in order to establish (2.3). To this end, suppose $R > 1$ is fixed, and consider the function

$$h(w, \phi) := \log |\psi(w) - \psi(e^{i\phi})| \quad \text{on} \quad \{w\colon |w| = R\} \times [0, 2\pi].$$

It is uniformly continuous there. Consequently, if we write $\phi_k = 2\pi k/(n+1)$ $(k = 0, 1, \ldots, n)$ and partition $[0, 2\pi]$ into the subintervals $i_k = [\phi_k, \phi_{k+1}]$ of length $2\pi/(n+1)$, then for each $\epsilon > 0$ the inequalities

$$h(w, \phi_k) \leqslant \min_{\phi \in i_k} h(w, \phi) + \epsilon \quad (k = 0, 1, \ldots, n)$$

hold for all w such that $|w| = R$, as soon as $n > N(\epsilon)$. Summation with respect to k yields

$$\frac{2\pi}{n+1} \Sigma_{k=0}^{n} h(w, \phi_k) \leqslant \frac{2\pi}{n+1} \Sigma_{k=0}^{n} \min_{\phi \in i_k} h(w, \phi) + 2\pi\epsilon,$$

and the first expression on the right-hand side can be regarded as a Riemann sum of an integral. We transform the left-hand side using $z = \psi(w)$ and $z_k^{(n)} = \psi(e^{i\phi_k})$, and we obtain

$$\text{(2.6)} \qquad \frac{1}{n+1} \log |\omega_n(z)| \leqslant \frac{1}{2\pi} \int_0^{2\pi} h(w, \phi) d\phi + \epsilon,$$

which is valid for all $z \in C_R$ and $n > N$.

The integral in (2.6) equals $2\pi \log (cR)$. To see this, we write

$$\int_0^{2\pi} = \int_0^{2\pi} \log \left| \frac{\psi(w) - \psi(e^{i\phi})}{w - e^{i\phi}} \right| d\phi + \int_0^{2\pi} \log |w - e^{i\phi}| d\phi.$$

According to the mean-value theorem for harmonic functions, the second integral equals $2\pi \log |w| = 2\pi \log R$, and the first integral equals

$$\text{Re} \left\{ \frac{1}{i} \int_{|t|=1} \log \frac{\psi(w) - \psi(t)}{w - t} \frac{dt}{t} \right\} \qquad (t = e^{i\phi}).$$

For fixed w ($|w| = R > 1$) the integrand is analytic in t for $|t| > 1$ and continuous for $|t| \geqslant 1$; its expansion at ∞ is of the form

$$\frac{\log c}{t} + \frac{1}{t^2} (\ldots),$$

so that the last integral equals $2\pi i \log c$.

Thus (2.6) yields

$$\frac{1}{n+1}\log|\omega_n(z)| \leqslant \log(cR) + \epsilon \qquad (z \in C_R;\, n > N).$$

This, in turn, implies the inequality

$$\frac{1}{n+1}\log M_n \leqslant \log(cR) + \epsilon \quad (n > N),$$

that is,

$$\limsup \sqrt[n+1]{M_n} \leqslant e^{\epsilon} cR.$$

Since $R > 1$ and $\epsilon > 0$ were arbitrary, it follows that $\limsup \sqrt[n+1]{M_n} \leqslant c$. Theorem 3 is therefore established.

Note that for the definition of the Fejér points the *exterior* mapping ψ was used. The interior mapping does not give uniformly distributed points in general, even for simple compact sets K. For example, if K is a disk $\{z\colon |z + r| \leqslant 1\}$ for some r $(0 < r < 1)$, the images $z_k^{(n)}$ of $w_k^{(n)} = e^{2\pi ik/(n+1)}$ under the mapping ϕ from $\mathbb{D}$ to K^o, with $\phi(0) = 0$, will not be uniformly distributed. The reader will find without difficulty that in this case $\sqrt[n+1]{M_n} \to 1 + r (n \to \infty)$.

Example. If $K = [-1, +1]$, then $\psi(w) = \frac{1}{2}(w + 1/w)$ furnishes the mapping from $\{w\colon |w| > 1\}$ onto K^c, and thus the Fejér points of order n are

$$z_k^{(n)} = \cos(2\pi k/(n+1)) \qquad (k = 0, 1, \ldots, n). \tag{2.7}$$

Some of these points coincide:

$$z_1^{(n)} = z_n^{(n)}, \quad z_2^{(n)} = z_{n-1}^{(n)}, \ldots,$$

and at these points f and f' must be interpolated. Our general results imply: Suppose E_R denotes the ellipse with semiaxes $a = \frac{1}{2}(R + 1/R)$ and $b = \frac{1}{2}(R - 1/R)$, where $R > 1$. If f is analytic inside and on E_R, then

$$|f(x) - L_n(x)| = O(1)R^{-n} \qquad (x \in [-1, +1];\, n \to \infty),$$

where the interpolating polynomials L_n are constructed from the Fejér points (2.7). This same rate of convergence was exhibited by the interpolating polynomials for the Chebyshev points; see §1, B.

D. The system of Fekete points

Here we become acquainted with another system of uniformly distributed points that was introduced by Fekete [1926]. We define these Fekete points for an *arbitrary* compact set $K \subset \mathbb{C}$ containing infinitely many points. For $z_k \in K$ $(k = 0, 1, \ldots, n)$ we take the product of the distances

$$P(z_0, z_1, \ldots, z_n) = \Pi_{j \neq k} |z_j - z_k|.$$

This is a bounded, continuous function of the $n + 1$ points on K. If P assumes its maximum for the choice $z_0^{(n)}, z_1^{(n)}, \ldots, z_n^{(n)}$, we call these points a system of *Fekete points of order n*. Obviously, all such points are distinct and "as far as possible apart from each other" on K. The following lemma contains their most important property.

Lemma 2. *For each system of Fekete points $z_k^{(n)}$ we have*

$$\text{(2.8)} \qquad \left| \frac{\omega_n(z)}{z - z_k^{(n)}} \right| \leqslant |\omega_n' (z_k^{(n)})| \qquad (k = 0, 1, \ldots, n; z \in K);$$

that is, the corresponding basic polynomials $l_k^{(n)}$ satisfy the condition

$$\text{(2.9)} \qquad |l_k^{(n)}(z)| \leqslant 1 \qquad (z \in K).$$

Proof. In order to show that (2.8) holds, say, for $k = 0$, we consider the function

$$F(z) := P(z, z_1^{(n)}, \ldots, z_n^{(n)}) = C \cdot \Pi_{k=1}^{n} |z - z_k^{(n)}|^2.$$

According to the definition of Fekete points, $F(z)$ assumes its maximum on K for $z = z_0^{(n)}$. Hence

$$\Pi_{k=1}^{n} |z - z_k^{(n)}| \leqslant \Pi_{k=1}^{n} |z_0^{(n)} - z_k^{(n)}| \qquad (z \in K),$$

and this is exactly (2.8) for $k = 0$.

If L_n denotes the interpolating polynomial corresponding to the Fekete points, property (2.9) now allows us to derive an estimate for the expression

$$\|f - L_n\| = \max_{z \in K} |f(z) - L_n(z)|,$$

provided an estimate of $\|f - P_n\|$ is known for *some arbitrary* polynomials P_n of degree n. The reason is that

$$f(z) - L_n(z, f) = [f(z) - P_n(z)] - L_n(z, f - P_n),$$

where

$$|L_n(z, f - P_n)| = |\Sigma_{k=0}^n [f(z_k^{(n)}) - P_n(z_k^{(n)})] l_k^{(n)}(z)| \leqslant \|f - P_n\| \cdot \Sigma_{k=0}^n |l_k^{(n)}(z)|$$
$$\leqslant (n+1) \|f - P_n\|$$

because of (2.9). It follows that

$$\|f - L_n\| \leqslant (n+2) \|f - P_n\|. \tag{2.10}$$

We now use the Fejér polynomials for such an estimate of $\|f - L_n\|$.

Theorem 4. *Suppose K is a compact subset of $\mathbb{C}$ and K^c is simply connected, suppose further that f is analytic on K. Then the interpolating polynomials L_n corresponding to the Fekete points satisfy the condition*

$$L_n(z) \rightrightarrows f(z) \qquad (n \to \infty; z \in K).$$

Combining this with the general Theorems 1 and 2, we obtain the following corollaries.

Corollary 1. *The Fekete points are uniformly distributed on K.*

Corollary 2. *On K, the L_n converge even maximally to f.*

Proof of Theorem 4. Since ψ might not be continuous on $\{w: |w| \geqslant 1\}$, we look for the Fejér points on a level curve C_R. Here $R > 1$ is so small that f is analytic inside and on C_R. Let P_n denote the interpolating polynomials corresponding to these points. In view of Theorems 2 and 3 they satisfy

$$|f(z) - P_n(z)| \leqslant M q^n \qquad (n = 0, 1, \ldots; z \in C_R)$$

for some $q < 1$. The same inequality holds in int $C_R \supset K$, and (2.10) now gives the desired result.

The Fekete points play an important role in determining and estimating the capacity of K; see Pommerenke [1975, Chapter 11]. Their distribution on ∂K can be studied fairly closely if ∂K is a sufficiently smooth Jordan curve; see Pommerenke [1967], [1969] and Kövari [1971]. There numerical experiments to determine the Fekete points are also reported.

Remarks about §2

1. Besides the systems of Fejér and Fekete, further systems of points are also important for complex interpolation, conformal mapping, and the solution of the Dirichlet problem. We mention here the point system of Leja [1957] that is obtained recursively, the extremal system of Menke [1972], [1974a], [1974b], [1975], [1976] that works with intermediate points, and the Curtiss points [1962], [1964], [1966], [1969a] (see also Siciak [1965] and Menke [1977]), which play a role in the interpolation by harmonic polynomials.

2. In the convergence theorems in Sections B and C, the behavior of $\sqrt[n]{|\omega_n(z)|}$ as $n \to \infty$ was crucial. The sequence $\{\omega_n(z)\}$ itself is examined under more detailed assumptions about ∂K by Curtiss [1941a], [1941b], who uses more precise statements about the convergence of Riemann sums.

3. In the theorems of §2, the domain of analyticity of f always contained the interpolation points in order that Hermite's formula could be used. If G is a Jordan domain, f is analytic in G and continuous in $\overline{G}$, and if one interpolates *on* ∂G, then there are two possible ways to examine the sequence $\{L_n\}$.

a) One uses Lagrange's formula (1.2); then one needs to examine closely the behavior of the basic polynomials $l_k^{(n)}$. See, for example, Curtiss [1935] or Gaier [1954].

b) One stays with Hermite's formula (1.3) and integrates across the nodes. Then one needs to interpret the integral in the sense of a principal value; see Curtiss [1969b].

4. For interpolation on the boundary of the domain of analyticity of f, the relation $L_n(z) \to f(z)$ $(n \to \infty; z \in \partial G)$ need not hold; see §4. However, under suitable assumptions about ∂G and f, we may have convergence in the mean:

$$\int_{\partial G} |f(z) - L_n(z)|^p \, |dz| \to 0 \qquad (n \to \infty);$$

see Al'per and Kalinogorskaja [1969].

5. Under some circumstances it is useful to employ polynomials of degree somewhat higher than n for the interpolation in $n + 1$ points. For example, Kövari [1968] proves the following. Suppose G is a Jordan domain with sufficiently smooth boundary; suppose f is analytic in G and continuous in $\overline{G}$; and suppose the nodes $z_k^{(n)} \in \partial G$ satisfy

$$|l_k^{(n)}(z)| \leqslant M \qquad (z \in \overline{G}; k = 0, 1, \ldots, n; n \geqslant 1).$$

Then, for $\eta > 0$, there exist polynomials P_n such that

a) $\deg P_n \leqslant n(1+\eta)$,

b) $P_n(z_k^{(n)}) = f(z_k^{(n)})$,

c) $P_n(z) \rightrightarrows f(z) \quad (n \to \infty; z \in \overline{G})$,

d) $P_n = P_n(\circ, f)$ is a linear operator in f.

§3. Approximation on more general compact sets; Runge's theorem

Up to now we have assumed that the complement of the compact set K, where we are to approximate by interpolation, is a simply connected domain; we have then been able to work conveniently with the conformal mapping ψ from $\{w: |w| > 1\}$ onto K^c. Now we admit a largely arbitrary compact set K, interpolate in the Fekete points of K, and prove independently from §2 the convergence of the interpolating polynomials, assuming f is analytic on K. A proof of Runge's approximation theorem results as an application.

A. Again: Interpolation in Fekete points

Suppose the compact set $K \subset \mathbb{C}$ is such that $K^c = \mathbb{C} \backslash K$ is a domain that has a Green's function G with pole at ∞. The latter is characterized by the following properties:

a) G is harmonic in K^c and is positive there;

b) $G(z) - \log |z|$ is bounded for $|z| \to \infty$;

c) $G(z) \to 0$ for $z \to \partial K$.

The existence of such a Green's function is assured if and only if K has positive capacity (see, for example, Goluzin [1969, p. 309]). Sufficient for the existence of G is that K consists of finitely many components (each of which contains more than one point).

The set

$$C_\rho = \{z: G(z) = \log \rho\} \qquad (\rho > 1)$$

is called a *level curve* corresponding to the parameter ρ. For all $\rho \neq \rho_k$ it consists of finitely many analytic Jordan curves γ_j such that

$$\gamma_j \cap \gamma_k = \emptyset, \quad \operatorname{int} \gamma_j \cap \operatorname{int} \gamma_k = \emptyset \quad (j \neq k),$$

and

$$K \subset \cup_j \operatorname{int} \gamma_j.$$

Here the ρ_k are at most countably many exceptional values with $\rho_k \to 1$, for which C_ρ passes through "critical points" of G, where finitely many γ_j meet (Walsh [1969, p. 67]). The set of exceptional values ρ_k is empty if and only if K^c is simply connected; this is the case we have considered so far. For increasing ρ, the domains bounded by the γ_j expand monotonically in the obvious way, and for each point $z \in K^c$ there exists exactly one ρ such that $z \in C_\rho$, namely $\rho = e^{G(z)}$.

Now if f is analytic on K (and not an entire function), then there exists a maximal ρ with the property that f has a unique analytic continuation from K to int C_ρ. Note here that in the various components of K completely different analytic functions can be defined, each of which can be continued to the corresponding part of int C_ρ. Consequently, C_ρ does not necessarily contain a singularity of the continuation of f; in fact, C_ρ need not contain such a singularity if it contains a critical point Q of G. In this case two analytic continuations meet at Q.

This maximal ρ now plays a crucial role for the rate of convergence of the interpolating polynomials.

Theorem 1. *Suppose $\rho > 1$ is the largest number such that f is analytic inside C_ρ, and suppose the L_n are the interpolating polynomials constructed with the Fekete points on K. Then*

$$\overline{\lim} \sqrt[n]{\max \{|f(z) - L_n(z)| : z \in K\}} = 1/\rho, \tag{3.1}$$

that is, the sequence $\{L_n\}$ converges maximally to f on K.

Note the remark at the end of §3.

Proof. a) First we observe that (3.1) cannot have "<". If this were the case, the relation

$$|f(z) - L_n(z)| \leq M/R^n,$$

and thus $|L_{n+1}(z) - L_n(z)| \leq 2M/R^n$ $(z \in K)$ would hold for the polynomials L_n and some $R > \rho$. An extension of Bernstein's lemma (Chapter I, §4) to the case where K^c might be multiply connected then shows

$$|L_{n+1}(z) - L_n(z)| \leq 2MR'(R'/R)^n$$

for all z inside the level curve $C_{R'}$. For $\rho < R' < R$ one finds that $L_n(z) \Rightarrow F(z)$, and f would have an analytic continuation F into the interior of $C_{R'}$; this contradicts the definition of ρ.

b) In order to show that $\overline{\lim} \leqslant 1/\rho$, we choose R_1 and R such that $1 < R_1 < R < \rho$; here we assume that R is not an exceptional value ρ_k, so that C_R consists of finitely many analytic Jordan curves.

Step 1: Suppose $d_1 \leqslant |t - z| \leqslant d_2$ $(z \in K, t \in C_{R_1})$; then

$$|\omega_n(z)/\omega_n(t)| \leqslant (n+2)d_2/d_1 \qquad (n = 1, 2, \ldots). \tag{3.2}$$

To see this, observe that the interpolating polynomial $L_n(z, f_0)$ corresponding to the function $f_0(z) = 1/(t - z)$ (where $t \in C_{R_1}$ is fixed) has the form

$$L_n(z, f_0) = \frac{1}{t - z} - \frac{\omega_n(z)}{\omega_n(t)(t - z)},$$

as already mentioned in §2, B; hence

$$|\omega_n(z)/\omega_n(t)| = |1 + (z - t)L_n(z, f_0)| \leqslant 1 + d_2 \cdot (n+1)/d_1 < (n+2)d_2/d_1.$$

Here we have used property (2.9) of the basic polynomials, namely $|l_k^{(n)}(z)| \leqslant 1$ $(z \in K)$, which is valid when the Fekete points are chosen as interpolation points.

Step 2: Next we find an estimate for the quotient (3.2) for $t \in C_R$:

$$\left|\frac{\omega_n(z)}{\omega_n(t)}\right| \leqslant (n+2)\frac{d_2}{d_1}\left(\frac{R_1}{R}\right)^{n+1} \qquad (z \in K; t \in C_R). \tag{3.3}$$

This is valid because (3.2) implies that

$$\log|\omega_n(t)| \geqslant \log\frac{d_1|\omega_n(z)|}{d_2(n+2)} =: A \qquad (z \in K \text{ fixed}; t \in C_{R_1});$$

hence we have for $t \in C_{R_1}$ that

$$h(t) := \log|\omega_n(t)| - (n+1)G(t) \geqslant A - (n+1)G(t) \geqslant A - (n+1)\log R_1.$$

On and outside C_{R_1}, the function $h(t)$ is harmonic and bounded for $t \to \infty$, so that the minimum principle can be applied:

$$\min\{h(t) : t \in C_R\} \geqslant \min\{h(t) : t \in C_{R_1}\};$$

consequently

$$\log|\omega_n(t)| - (n+1)\log R \geqslant A - (n+1)\log R_1 \qquad (t \in C_R),$$

and this is (3.3).

Step 3: Estimate of $f - L_n$. According to Hermite's formula, we have

$$f(z) - L_n(z) = \frac{1}{2\pi i}\int_{C_R} \frac{\omega_n(z)}{\omega_n(t)}\,\frac{f(t)}{t-z}\,dt \qquad (z \in K),$$

and taking (3.3) into account, we see that

$$f(z) - L_n(z) = O(1)\cdot(n+2)(R_1/R)^n \qquad (z \in K; n \to \infty).$$

Hence

$$\overline{\lim}\ \sqrt[n]{\max\{|f(z) - L_n(z)| : z \in K\}} \leqslant R_1/R,$$

and since $R_1 > 1$ and $R < \rho$ were arbitrary, we have that $\overline{\lim} \leqslant 1/\rho$.

B. Runge's approximation theorem

Theorem 1 offers an opportunity to prove the important theorem of Runge.

Theorem 2 (Runge 1885). *Suppose $K \subset \mathbb{C}$ is compact, $K^c = \mathbb{C}\backslash K$ is connected, and f is analytic on K. Then there exist polynomials P_n such that*

$$\max\{|f(z) - P_n(z)| : z \in K\} \to 0 \qquad (n \to \infty).$$

This theorem marks the beginning of complex approximation theory. It was published in the same year as the theorem of Weierstrass about the approximation of continuous functions on intervals. Runge conducts the proof of Theorem 2 using *rational* approximation followed by shifting of the poles; we shall return to this again in Chapter III, §1.

Proof. Since K^c is a domain, it can be exhausted by a sequence of finitely connected domains (Walsh [1969, pp. 8-9]); this means: There are domains G_j such that

i) $\overline{G}_j \subset G_{j+1} \subset K^c$;

ii) $\cup G_j = K^c$;

iii) $\partial G_j = \Gamma_1^{(j)} \cup \Gamma_2^{(j)} \cup \ldots \cup \Gamma_{N_j}^{(j)}$ with rectifiable Jordan curves Γ.

Since f is analytic on K, there is an index j such that f is analytic on and inside the components of ∂G_j; we fix this index and let k_i denote the closure of the domain int $\Gamma_i^{(j)}$ $(i = 1, 2, \ldots, N_j)$.

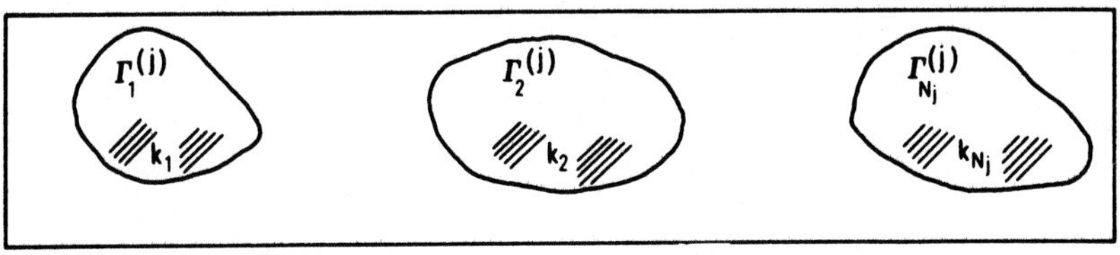

Now f is analytic on the compact set $k := \cup_{i=1}^{N_j} k_i$ whose complement has a Green's function. This allows us to apply Theorem 1, and we conclude that the interpolating polynomials P_n that are constructed for the Fekete points on k satisfy an estimate

$$\max \{|f(z) - P_n(z)| : z \in k\} = O(q^n) \qquad (n \to \infty)$$

for some $q < 1$. Since $k \supset K$, the assertion follows.

Next we analyze the assumptions of Theorem 2.

1. *If K^c is not connected, the theorem is false.* For if K^c is not connected and thus has a bounded component g, we choose $z_0 \in g$, write $d =$ $\max \{|z - z_0| : z \in K\}$, and consider the function $f(z) = 1/(z - z_0)$, which is analytic on K. If there were a polynomial P such that

$$|P(z) - f(z)| < 1/d \qquad (z \in K),$$

then

$$|P(z)(z - z_0) - 1| < \frac{|z - z_0|}{d} \leqslant 1 \qquad (z \in K).$$

But $\overline{g}$ is compact, and the polynomial $P(z)(z - z_0) - 1$ assumes its maximum on $\overline{g}$ at the boundary $\partial g \subset K$. This implies that

$$|P(z)(z - z_0) - 1| < 1 \qquad (z \in g),$$

which obviously is false for $z_0 \in g$.

2. The requirement that f be analytic on K can, however, be weakened. In order for the conclusion of Theorem 2 to remain valid, it is sufficient that f is continuous on K and analytic at the interior points of K (if there are any). It is considerably more difficult to prove this theorem of Mergelyan; we shall return to it in Chapter III, §2.

Remark about § 3

In this generality, Theorem 1 was first proved by Walsh and Russell [1934]. The proof presented here goes back to Shen [1936, pp. 158-159]; see also Walsh [1969, p. 173]. The proof can be adapted to interpolation by rational functions. To this end, let $\alpha_k^{(n)}$ $(k = 0, 1, \ldots, n)$ denote prescribed poles, which may only have accumulation points in the exterior of C_ρ, and choose the interpolation points $z_k^{(n)}$ on K such that

$$\Pi_{j<k} |z_j^{(n)} - z_k^{(n)}| / \Pi_{j,k=0}^{n} |z_j^{(n)} - \alpha_k^{(n)}|$$

becomes a maximum. Instead of polynomials, Shen then allows rational functions of the form

$$\frac{a_0 + a_1 z + \ldots + a_n z^n}{(z - \alpha_0^{(n)})(z - \alpha_1^{(n)}) \ldots (z - \alpha_n^{(n)})}$$

for the interpolation, and he then proves a theorem that generalizes Theorem 1.

§4. Interpolation in the unit disk

We shall study this special case more closely, because here several more subtle statements can be made about the convergence and divergence of interpolating polynomials. In addition, we present a theorem about approximation in the unit disk by rational functions, where the interpolation points also lie in the unit disk.

A. Interpolation on $\{z: |z| = r\}, r < 1$

The situation is the clearest, if we choose the interpolation points

$$z_k^{(n)} = re^{2\pi ik/(n+1)} \qquad (k = 0, 1, \ldots, n; 0 < r < 1). \tag{4.1}$$

Suppose the function f, which is to be approximated, is analytic in $\mathbb{D} = \{z: |z| < 1\}$; let L_n denote the nth interpolating polynomial corresponding to the points $z_k^{(n)}$, and let S_n denote the nth partial sum of the Taylor series of f at 0. As in §1, B, Hermite's interpolation formula now leads to the representation

$$L_n(z) - S_n(z) = r^{n+1} \cdot \frac{1}{2\pi i} \int_{|t|=\tau} \frac{t^{n+1} - z^{n+1}}{t^{n+1}(t-z)(t^{n+1} - r^{n+1})} f(t)dt \qquad (z \in \mathbb{C}),$$

where $r < \tau < 1$. For $|z| = R > 1$ one finds that $L_n(z) - S_n(z) = O(r^n R^n / \tau^{2n})$ $(n \to \infty)$, and since τ may be arbitrarily close to 1, we see that

$$(4.2) \qquad L_n(z) - S_n(z) \rightrightarrows 0 \qquad (n \to \infty;\ |z| \leqslant R)$$

for each $R < 1/r$. *Thus the sequences* $\{L_n\}$ *and* $\{S_n\}$ *are equiconvergent in a disk containing* $\mathbb{D}$.

We draw the following *conclusions* from this:

1) In each compact subset of $\mathbb{D}$ we have that $L_n(z) \rightrightarrows f(z)$ $(n \to \infty)$.

2) There exists a function f analytic in $\mathbb{D}$ and continuous in $\overline{\mathbb{D}}$ whose interpolating polynomials L_n corresponding to (4.1) are unbounded at $z = 1$: $\sup_n |L_n(1)| = \infty$; for, as is well known (Fejér, 1910), there exists such an f with $\sup_n |S_n(1)| = \infty$.

3) On the other hand, also according to Fejér (1904), the arithmetic means of the S_n approach f for $|z| = 1$, provided f is continuous in $\overline{\mathbb{D}}$ and analytic in $\mathbb{D}$. For such f we thus have

$$\frac{1}{n+1} \Sigma_{k=0}^{n} L_k(z) \rightrightarrows f(z) \qquad (n \to \infty; z \in \overline{\mathbb{D}}).$$

Statement (4.2) will now be examined more closely.

First, the domain of validity $\{z: |z| \leqslant R\}$ $(R < 1/r)$ can, in general, not be enlarged. In particular, if one takes $f(z) = 1/(z-1)$, then

$$L_n(z) = \frac{1}{z-1} - \frac{z^{n+1} - r^{n+1}}{(1 - r^{n+1})(z-1)}$$

and

$$S_n(z) = \frac{1 - z^{n+1}}{z - 1},$$

so that

$$L_n(z) - S_n(z) = \frac{r^{n+1}}{1 - r^{n+1}} \cdot \frac{1 - z^{n+1}}{z - 1}.$$

For $z = 1/r$, the right-hand side becomes $-r/(1 - r)$ and thus does not tend to zero.

But it is possible to enlarge the domain of validity by using different polynomials of degree n instead of the partial sums S_n of the Taylor series $f(z) = \Sigma_{k=0}^{\infty} a_k z^k$ ($z \in \mathbb{D}$). Cavaretta, Sharma, and Varga [1980] have pointed this out in a recent work. (There everything is in the context of interpolation in the roots of unity, and f is assumed to be analytic in $\{z: |z| < \rho\}$ for some $\rho > 1$.) To see this, we consider the polynomials of degree n

$$S_{n,j}(z) := \Sigma_{k=0}^{n} a_{j(n+1)+k} z^k \qquad (j = 0, 1, 2, \ldots);$$

obviously, $S_{n,0}(z) = S_n(z)$. They have the integral representation

$$S_{n,j}(z) = \frac{1}{2\pi i} \int_{|t|=\tau} \frac{t^{n+1} - z^{n+1}}{t - z} \cdot \frac{f(t)}{t^{(j+1)(n+1)}}\, dt,$$

and for the polynomials $\Sigma_{j=0}^{l} S_{n,j}(z) r^{j(n+1)}$ ($l = 0, 1, 2, \ldots$), also of degree n, one finds that

$$\Sigma_{j=0}^{l} S_{n,j}(z) r^{j(n+1)} = \frac{1}{2\pi i} \int_{|t|=\tau} \frac{t^{n+1} - z^{n+1}}{t - z} \cdot \frac{t^{(l+1)(n+1)} - r^{(l+1)(n+1)}}{(t^{n+1} - r^{n+1}) t^{(l+1)(n+1)}} f(t) dt.$$

Hence

$$L_n(z) - \Sigma_{j=0}^{l} S_{n,j}(z) r^{j(n+1)} =$$

$$r^{(l+1)(n+1)} \cdot \frac{1}{2\pi i} \int_{|t|=\tau} \frac{t^{n+1} - z^{n+1}}{t - z} \cdot \frac{f(t)dt}{(t^{n+1} - r^{n+1}) t^{(l+1)(n+1)}},$$

which is a generalization of the formula following (4.1) above. For $|z| = R > 1$, the right-hand side is $O(r^{(l+1)n} R^n / \tau^{(l+2)n})$ ($n \to \infty$), and since τ may be arbitrarily close to 1, we obtain

$$(4.2') \qquad L_n(z) - \Sigma_{j=0}^{l} S_{n,j}(z) r^{j(n+1)} \rightrightarrows 0 \qquad (n \to \infty; |z| \leq R)$$

as long as $R < 1/r^{l+1}$. For $l = 0$ this becomes (4.2).

The bound r^{-l-1} is again sharp, but (4.2′) holds even for $|z| \leq r^{-l-1}$ if f has a continuous extension to $\bar{\mathbb{D}}$ or at least an integrable boundary function. For details we refer to the cited work by Cavaretta, Sharma, and Varga [1980]. There corresponding theorems about Hermite interpolation and Birkhoff-Hermite interpolation are also proved.

B. Interpolation on $\{z: |z| = 1\}$

If one takes the $(n+1)^{\text{st}}$ roots of unity $z_k^{(n)} = e^{2\pi ik/(n+1)}$ $(k = 0, 1, \ldots, n)$ as interpolation points, one must refer back to Lagrange's formula. We have $\omega(z) = z^{n+1} - 1$ and $\omega'(z) = (n+1)z^n = \dfrac{n+1}{z}$ if $z = z_k^{(n)}$; hence

$$L_n(z) = (1 - z^{n+1}) \cdot \frac{1}{2\pi} \Sigma_{k=0}^{n} \frac{f(z_k^{(n)}) z_k^{(n)}}{z_k^{(n)} - z} \frac{2\pi}{n+1}. \tag{4.3}$$

We recognize the second factor as a Riemann sum for $\dfrac{1}{2\pi} \displaystyle\int_0^{2\pi} \frac{f(e^{i\phi})e^{i\phi}}{e^{i\phi} - z} \, d\phi$, and we obtain the result: *If f is continuous on* $\{z: |z| = 1\}$, *then*

$$L_n(z) \to \frac{1}{2\pi i} \int_{\partial\mathbb{D}} \frac{f(\zeta)}{\zeta - z} \, d\zeta \qquad (n \to \infty; z \in \mathbb{D})$$

uniformly on compact subsets of $\mathbb{D}$. The result can be extended to (smooth) Jordan curves; see Curtiss [1935]. In particular, if $f \in A$ (i.e., f is continuous in $\bar{\mathbb{D}}$ and analytic in $\mathbb{D}$), then $L_n(z) \rightrightarrows f(z)$ $(n \to \infty)$ on each compact subset of $\mathbb{D}$.

But what is the behavior of $\{L_n\}$ *on the boundary of the unit disk?* We now turn to this question.

We begin with a positive result. The numbers

$$E_n = \inf \max \{|f(z) - P(z)|: z \in \bar{\mathbb{D}}\}$$

enter into this, where the infimum is taken over all polynomials P of degree n.

Theorem 1. *Suppose $f \in A$, that is, f is continuous in $\bar{\mathbb{D}}$ and analytic in $\mathbb{D}$, and suppose L_n is the interpolating polynomial* (4.3). *Then*

$$|f(z) - L_n(z)| \leq M E_n \log(n+1) \qquad (n = 1, 2, \ldots; z \in \bar{\mathbb{D}}), \tag{4.4}$$

where M is some absolute constant.

For the proof we need the following lemma.

Lemma 1. *Suppose P is a polynomial of degree $n \geqslant 1$, and suppose $|P(z_k^{(n)})| \leqslant 1$ for the $(n+1)^{\text{st}}$ roots of unity $z_k^{(n)}$ $(k = 0, 1, \ldots, n)$. Then*

$$|P(z)| < 3 \log (n+1) \quad \textit{for} \quad |z| \leqslant 1. \tag{4.5}$$

The order of magnitude of the right-hand side is best possible, even though 3 could be replaced by a smaller constant. For this topic see Gronwall [1920].

Proof. For $n = 1$ we write P as

$$P(z) = \frac{1+z}{2} P(1) + \frac{1-z}{2} P(-1),$$

which implies that $|P(z)| \leqslant \frac{1}{2}[|1+z| + |1-z|] \leqslant \sqrt{2}$ if $|z| \leqslant 1$. Hence (4.5) holds for $n = 1$.

For $n \geqslant 2$, we begin with the representation of P by Lagrange's formula (see (4.3)) and obtain

$$|P(z)| \leqslant \frac{1}{n+1} \Sigma_{k=0}^{n} \frac{|1 - z^{n+1}|}{|z_k^{(n)} - z|}.$$

We can now assume that $|z| = 1$ and $0 < \arg z < \alpha := \dfrac{2\pi}{n+1}$. In the last sum each summand is bounded by $n + 1$, because

$$\left| \frac{1 - z^{n+1}}{z_k^{(n)} - z} \right| = |(z_k^{(n)})^n + (z_k^{(n)})^{n-1} z + \ldots + (z_k^{(n)}) z^{n-1} + z^n| \leqslant n + 1. \tag{4.6}$$

We estimate the terms corresponding to $k = 0$ and $k = 1$ by $n + 1$. For the *remaining* $z_k^{(n)}$ on each side of the diameter of the unit disk that passes through 0 and z, the arc between z and $z_k^{(n)}$ has length $\geqslant \alpha, 2\alpha, 3\alpha, \ldots, p\alpha$, respectively, where $p\alpha \leqslant \pi$. Hence $|z_k^{(n)} - z|$ is at least $2 \sin \alpha/2, 2 \sin 2\alpha/2, \ldots, 2 \sin p\alpha/2$, and since $\sin x \geqslant (2/\pi)x$ for $0 \leqslant x \leqslant \pi/2$, it is at least $2\alpha/\pi$, $4\alpha/\pi, \ldots, 2p\alpha/\pi$, respectively. For these $z_k^{(n)}$ we thus have that

$$\left| \frac{1 - z^{n+1}}{z_k^{(n)} - z} \right| \quad \text{is at most} \quad \frac{\pi}{\alpha}, \frac{\pi}{2\alpha}, \ldots, \frac{\pi}{p\alpha}.$$

Hence

$$|P(z)| \leqslant 2 + \frac{2}{n+1} \cdot \frac{\pi}{\alpha} \Sigma_{j=1}^{p} \frac{1}{j} = 2 + \Sigma_{j=1}^{p} \frac{1}{j} < 3 + \log p \leqslant 3 + \log \frac{\pi}{\alpha},$$

so that

$$|P(z)| < 3 + \log \frac{n+1}{2} \quad \text{for} \quad |z| \leqslant 1.$$

For $n \geqslant 3$, the right-hand side is $< 3 \log (n+1)$, as asserted in (4.5). And for $n = 2$, (4.6) implies that $|P(z)| \leqslant 3$, and therefore (4.5) holds for all $n \geqslant 1$.

Proof of Theorem 1. Suppose P is a polynomial of degree n such that $\|f - P\| = E_n$. Then

$$f(z) - L_n(z, f) = [f(z) - P(z)] - L_n(z, f - P),$$

because $L_n(z, P) = P(z)$. From this and (4.5) we get for $n \geqslant 1$:

$$\|f - L_n\| \leqslant \|f - P\| + \|L_n(\cdot, f - P)\| \leqslant E_n + 3\, E_n \log (n+1),$$

since $L_n(\cdot, f - P)$ is a polynomial of degree n whose values at the $(n+1)^{\text{st}}$ roots of unity are bounded in absolute value by E_n. This establishes (4.4).

Under the assumptions of Theorem 1, the E_n always tend to zero, but $L_n(z) \rightrightarrows f(z)$ $(n \to \infty;\, z \in \overline{\mathbb{D}})$ is assured only if even $E_n \log (n+1) \to 0$ $(n \to \infty)$. This is the case, for example, if $f \in \operatorname{Lip} \alpha$ on $\partial \mathbb{D}$ for some $\alpha > 0$; see Chapter I, §6D. But in general the sequence $\{L_n(-1)\}$ can be unbounded, as Fejér [1916] has shown with a tricky construction.

For the sake of technical simplification we interpolate in the points

$$w_k^{(n)} = -z_k^{(n)} = -e^{2\pi i k/(n+1)} \qquad (k = 0, 1, \ldots, n).$$

We let U_n denote the interpolating polynomials. Then the following result holds.

Theorem 2. *There exists a function f, continuous in $\overline{\mathbb{D}}$ and analytic in $\mathbb{D}$, for which the sequence $\{U_n(1, f)\}$ is unbounded.*

Proof. Fejér works with the polynomials

$$h_p(z) := \left(\frac{1}{p} + \frac{z}{p-1} + \ldots + \frac{z^{p-1}}{1} \right) - \left(\frac{z^{p+1}}{1} + \frac{z^{p+2}}{2} + \ldots + \frac{z^{2p}}{p} \right)$$

$(p \in \mathbb{N})$, which he introduced earlier. First we need to study $U_n(1, h_p)$ for various values of p and n.

i) If n is odd, then 1 is an interpolation point, so that

$$U_n(1, h_p) = h_p(1) = 0 \quad \text{for all odd } n.$$

ii) If $n \geqslant 2p$, then $U_n(z, h_p) = h_p(z)$, so that

$$U_n(1, h_p) = h_p(1) = 0 \quad \text{for all } n \geqslant 2p.$$

iii) Now we study $U_n(1, h_p)$ for $p = n + 1, p = 3(n + 1), \ldots, p = u(n + 1)$, where n is even and u is odd.

For $p = n + 1$, we have $(w_k^{(n)})^p = (-1)^p = -1$, so that

$$h_p(w_k^{(n)}) = \left(\frac{1}{p} + \frac{w_k^{(n)}}{p-1} + \ldots + \frac{(w_k^{(n)})^{p-1}}{1}\right)$$

$$+ \left(\frac{w_k^{(n)}}{1} + \frac{(w_k^{(n)})^2}{2} + \ldots + \frac{(w_k^{(n)})^{p-1}}{p-1} - \frac{1}{p}\right) = g_1(w_k^{(n)}),$$

where

$$g_1(z) = \left(\frac{1}{p-1} + 1\right) z + \left(\frac{1}{p-2} + \frac{1}{2}\right) z^2 + \ldots + \left(\frac{1}{1} + \frac{1}{p-1}\right) z^{p-1}.$$

Hence $U_n(z, h_p) = g_1(z)$, and consequently

$$U_n(1, h_p) = g_1(1) = 2 \, \Sigma_{j=1}^{n} \, 1/j \quad \text{for } p = n + 1, \quad n \text{ even.}$$

For $p = 3(n + 1)$, one finds in the same manner that $h_p(w_k^{(n)}) = g_3(w_k^{(n)})$, where

$$g_3(z) = \left(\frac{1}{p-1} + 1\right) z + \left(\frac{1}{p-2} + \frac{1}{2}\right) z^2 + \ldots + \left(\frac{1}{p-n} + \frac{1}{n}\right) z^n$$

$$- \left[\left(\frac{1}{p-n-2} + \frac{1}{n+2}\right) z + \ldots + \left(\frac{1}{p-2n-1} + \frac{1}{2n+1}\right) z^n\right]$$

$$+ \left(\frac{1}{p-2n-3} + \frac{1}{2n+3}\right) z + \ldots + \left(\frac{1}{1} + \frac{1}{3n+2}\right) z^n.$$

Hence

$$U_n(z, h_p) = g_3(z), \text{ and therefore } U_n(1, h_p) = g_3(1),$$

and the latter obviously is positive. We note that

$$U_n(1,h_p) > 0 \quad \text{for } p = 3(n+1), \quad n \text{ even.}$$

Analogously one shows that in general

$$U_n(1,h_p) > 0 \quad \text{for } p = u(n+1), \quad n \text{ even}, \quad u \text{ odd.}$$

Now the desired f is defined by a series expansion

$$f(z) := \Sigma_{k=1}^{\infty} k^{-2} h_{p_k}(z), \text{ where } p_k = 3^{k^3}.$$

This series converges uniformly in $\bar{\mathbb{D}}$, since $|h_p(z)| \leqslant M$ for all $z \in \bar{\mathbb{D}}$ and $p \in \mathbb{N}$. Hence $f \in A$. In addition,

$$U_n(1,f) = \Sigma_{k=1}^{\infty} k^{-2} U_n(1,h_{p_k}).$$

We now write $n = 3^{k_0^3} - 1$ for some $k_0 \in \mathbb{N}$ and decompose U_n as follows:

$$U_n(1,f) = \Sigma_{k<k_0} k^{-2} U_n(1,h_{p_k}) + k_0^{-2} U_n(1,h_{p_{k_0}}) + \Sigma_{k>k_0} k^{-2} U_n(1,h_{p_k}).$$

In the first sum we have $p_k \leqslant n/2$, so that each summand is 0 by ii). In the last sum, $p_k = 3^{k^3} = u \cdot 3^{k_0^3} = u(n+1)$, where the factor u is odd; hence each summand is > 0 by iii). But in the middle expression we have $p_{k_0} = n+1$, where n is even; hence, by iii), the summand in the middle is

$$k_0^{-2} \cdot 2\,\Sigma_{j=1}^{n} 1/j > 2\,k_0^{-2} \log(n+1) = 2\,k_0^{-2} \cdot k_0^3 \log 3.$$

All together, the sequence $\{U_n(1,f)\}$ is unbounded; Theorem 2 is established.

In order to avoid the phenomenon of divergence expressed in Theorem 2, one could consider those polynomials H_n of degree $2n+1$ that satisfy

$$H_n(z_k^{(n)}) = f(z_k^{(n)}) \quad \text{and} \quad H_n'(z_k^{(n)}) = 0 \qquad (k = 0, 1, \ldots, n).$$

For each continuous function f and for a suitable choice of the interpolation points, this Hermite interpolation always leads to a convergent process in the real case; see, for example, Natanson [1955, p. 397]. However, one can show here that

$$H_n(z) = (1 - z^{n+1})L_n(z) + O(1) \qquad (n \to \infty; z \in \bar{\mathbb{D}})$$

(Losinsky [1939, p. 320]), so that the unboundedness of $\{H_n(-1)\}$ follows from that of $\{L_n(-1)\}$ (n even). By the way, this is true also if one requires more generally that

$$H_n(z_k^{(n)}) = f(z_k^{(n)}) \quad \text{and} \quad H_n'(z_k^{(n)}) = \alpha_k^{(n)},$$

where $\alpha_k^{(n)} = o(n/\log n)$ $(n \to \infty)$ uniformly in k; see Gaier [1954, p. 131].

C. Approximation by rational functions

From this rich subject we will select only one specific result, which is connected with the interpolation in the unit disk. Moreover, it has a certain practical significance.

Theorem 3. *Suppose* $f \in A$, *and suppose* $\alpha_1, \alpha_2, \ldots, \alpha_n$ *are pairwise distinct points in* $\{z\colon |z| > 1\}$. *For rational functions of the form*

$$r(z) = \frac{a_0 + a_1 z + \ldots + a_n z^n}{(z-\alpha_1)(z-\alpha_2)\ldots(z-\alpha_n)} \tag{4.7}$$

the expression

$$\int_{\partial \mathbb{D}} |f(z) - r(z)|^2 \, |dz|$$

is a minimum if and only if r interpolates the given function f at the points $0, 1/\bar{\alpha}_1, \ldots, 1/\bar{\alpha}_n$.

Proof. Suppose r^* is this interpolating rational function of the form (4.7); it is uniquely determined. We show that

$f - r^*$ is orthogonal to the functions 1 and $1/(z - \alpha_k)$ $(k = 1, 2, \ldots, n)$.

For we have

$$\int_{\partial \mathbb{D}} (f - r^*) \cdot 1 |dz| = \frac{1}{i} \int_{\partial \mathbb{D}} (f - r^*) \frac{dz}{z} = 0,$$

because $f - r^*$ vanishes at 0, and further

$$\int_{\partial\mathbb{D}} (f - r^*) \cdot \overline{\frac{1}{z - \alpha_k}} \, |dz| = \frac{1}{i} \int_{\partial\mathbb{D}} (f - r^*) \frac{dz}{1 - \bar{\alpha}_k z} = 0,$$

because $f - r^*$ vanishes at $1/\bar{\alpha}_k$.

Now we see that $\int_{\partial\mathbb{D}} |f - r|^2 \, |dz|$ is a minimum if and only if $r = r^*$. The reason is that

$$\begin{aligned} \int |f - r|^2 \, |dz| &= \int [(f - r^*) + (r^* - r)] \, [(f - r^*) + (r^* - r)]^- \, |dz| \\ &= \int |f - r^*|^2 \, |dz| + \int |r^* - r|^2 \, |dz| + 0; \end{aligned}$$

here we have used that

$$r^* - r = \Sigma_{k=1}^{n} \frac{A_k}{z - \alpha_k} + A_0,$$

where the A_k are constants. The last statement holds, because $r^* - r$ is a rational function with simple poles at $z = \alpha_k$, which remains bounded for $z \to \infty$. Everything follows from this.

If one takes more and more poles $\alpha_k^{(n)}$, one obtains a sequence $\{r_n\}$ of rational functions of best approximation for which the relation

$$r_n(z) \Rightarrow f(z) \qquad (n \to \infty; z \in \bar{\mathbb{D}})$$

holds if f is analytic in $\bar{\mathbb{D}}$ and if the poles do not have a limit point on $\partial\mathbb{D}$. For details, see Walsh [1969, p. 245].

Remarks about §4

1. The proof of Theorem 2 could be simplified somewhat by using the Banach-Steinhaus theorem. Consider the functionals $U_n(1, f)$ for even n and for f in the Banach space A. We saw in iii) that $|U_n(1, h_p)| > c \log n$ for certain polynomials h_p $(p = n + 1)$ with $\|h_p\| \leqslant M$. Hence the norms $\|U_n(1, \cdot)\|$ are unbounded, and the Banach-Steinhaus theorem guarantees the existence of an $f \in A$ with unbounded $U_n(1, f)$.

Divergence almost everywhere on $|z| = 1$ is discussed in a recent paper by German [1980].

2. Interpolation theory in the real domain often depends on summability processes. For example, if f has period 2π and is continuous on $\mathbb{R}$, and if

$$U_n(x, f) = \frac{a_0^{(n)}}{2} + \Sigma_{k=1}^{n} (a_k^{(n)} \cos kx + b_k^{(n)} \sin kx)$$

is the interpolating polynomial corresponding to f and the points $x_k^{(n)} = 2\pi k/(2n + 1)$ $(k = 0, 1, \ldots, 2n)$, one writes

$$U_{n,j}(x, f) = \frac{a_0^{(n)}}{2} + \Sigma_{k=1}^{j} (a_k^{(n)} \cos kx + b_k^{(n)} \sin kx)$$

and examines, for example, the arithmetic means

$$\sigma_n(x, f) = \frac{1}{n+1} \Sigma_{j=0}^{n} U_{n,j}(x, f) \tag{4.8}$$

for convergence. The general process of this type is defined by

$$\sigma_n(x, f) = \Sigma_{k=0}^{n} (a_k^{(n)} \cos kx + b_k^{(n)} \sin kx)\lambda_k^{(n)}$$

with certain weights $\lambda_k^{(n)}$; see, for example, Natanson [1955, p. 406 ff.] and Zygmund [1959, p. 22 ff.]. Note that in forming these means, only the values of f at the points $x_k^{(n)}$, for the respective n, are used.

Means of the form $\Sigma_{k=0}^{n} \lambda_k^{(n)} U_k(x, f)$, such as the C_1 - means

$$\tau_n(x, f) = \frac{1}{n+1} \Sigma_{k=0}^{n} U_k(x, f),$$

are discussed less often. Here interpolating polynomials corresponding to *different* systems of interpolation points are averaged. In this context Marcinkiewicz [1936, p. 5], for example, shows that there exist a continuous function f with period 2π and a point x_0 such that $\{\tau_n(x_0, f)\}$ does not remain bounded.

These methods could be used to study limits of the polynomials $L_n(z, f)$, which interpolate on $|z| = 1$; but little has been done. Berman [1971] applies a process analogous to (4.8) to $\{L_n\}$, and Gaier [1954] constructs an f for which not only $\{L_n(-1)\}$, but even the Borel means of this sequence are unbounded.

Quite recently Vértesi [1983] proved the following surprising result: Given any row-finite summability matrix M, there exists an $f \in A$ such that the M-means of the interpolating polynomials, formed for f and the roots of unity, are unbounded. In particular, *the Cesàro means*

$$A_n(z, f) := \frac{1}{n+1} \Sigma_{k=0}^{n} L_k(z, f)$$

may be unbounded for an $f \in A$. This is shown by proving that the norms $\|A_n\|$ diverge like $\log n$.

3. If f is analytic in $\overline{\mathbb{D}}$, then $L_n(z, f) \Rightarrow f(z)$ $(n \to \infty; z \in \overline{\mathbb{D}})$ is valid for interpolation in uniformly distributed points on $\partial\mathbb{D}$. Hlawka [1969] deals with the question of what approximation statement can be made, if the interpolation points are not uniformly distributed. Here a measure of the deviation from the uniform distribution plays a role.

PART II

GENERAL APPROXIMATION THEOREMS IN THE COMPLEX PLANE

In contrast to the treatment up to now, the constructive points of view will not play such an important role here. Instead we shall now deal with questions about the *existence* of approximating rational, entire, or meromorphic functions. At first the set, where the approximation is to take place, is a compact set K in $\mathbb{C}$; later, in Chapter IV, it is merely a closed set F in an arbitrary domain $G \subset \mathbb{C}$.

Chapter III

APPROXIMATION ON COMPACT SETS

For the approximation of a function f on a compact set K polynomials or, more generally, rational functions can be utilized. The situation is simple if f is analytic on K, whereas a weakening of this assumption requires a much greater effort. We begin with Runge's theorem; a weak version of it has already been proved, but the following proof will make Chapter III independent of the preceding chapters.

§1. Runge's approximation theorem

This theorem (proved in 1885) stands at the beginning of complex approximation. A function f is to be approximated by rational functions R, uniformly on a compact set $K \subset \mathbb{C}$. We have $R(z) = \dfrac{P(z)}{Q(z)}$, where P and Q are polynomials (without common factors), or

$$R(z) = \Sigma_j P_j\left(\frac{1}{z - z_j}\right) + P_0(z),$$

where P_0, P_j are finitely many polynomials (partial fraction decomposition). Compact sets can be very complicated; an example shall suffice (see also §3, A):

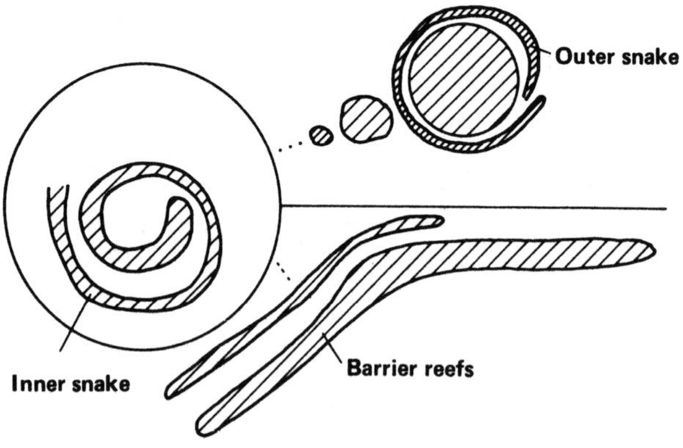

A. General Cauchy formula

We shall need the following result.

Theorem 1. *Suppose K is a compact set in $\mathbb{C}$ and $U \supset K$ is an open set. Then there exist finitely many horizontal or vertical, oriented line segments $\gamma_1, \gamma_2, \ldots, \gamma_N$ in $U \backslash K$ such that if f is analytic in U and if $\Gamma = \gamma_1 + \ldots + \gamma_N$, then*

$$f(z) = \frac{1}{2\pi i} \int_\Gamma \frac{f(\zeta)}{\zeta - z}\, d\zeta \qquad (z \in K).$$

Note that Γ depends on K and U, but not on f.

Proof. Let $\delta := \operatorname{dist}(K, \partial U) > 0$. We introduce a square grid of mesh h in the plane, where $h\sqrt{2} < \delta$. All closed squares of the grid that intersect K get numbered: $Q_1, Q_2, \ldots, Q_p$. Then

$$K \subset \cup Q_j \subset U.$$

The inclusion on the left is obvious, and if $z_0 \in Q_j \cap K$, then the entire disk about z_0 with radius δ is in U, and with it also the square Q_j of diameter $h\sqrt{2} < \delta$.

Now suppose the boundary of each Q_j has positive orientation. If two squares Q_j, Q_k have a common side γ, we discard it. Let $\gamma_1, \gamma_2, \ldots, \gamma_N$ denote the remaining oriented sides of the Q_j. No γ_j meets K; otherwise γ_j would belong to two adjacent squares. Hence

$$\Gamma := \gamma_1 + \gamma_2 + \ldots + \gamma_N \subset U \backslash K.$$

Suppose now that f is analytic in U and $z \in K$, so that $z \in Q_{j_0}$ for at least one j_0. If in addition $z \in Q^\circ_{j_0}$, the formula of Theorem 1 is valid:

$$f(z) = \frac{1}{2\pi i} \int_{\partial Q_{j_0}} \frac{f(\zeta)}{\zeta - z}\, d\zeta = \frac{1}{2\pi i} \Sigma_j \int_{\partial Q_j} \frac{f(\zeta)}{\zeta - z}\, d\zeta$$

$$= \frac{1}{2\pi i} \int_\Gamma \frac{f(\zeta)}{\zeta - z}\, d\zeta.$$

And if $z \in \partial Q_{j_0}$, then z belongs to two adjacent squares and thus lies on a discarded side; consequently $z \notin \Gamma$. A continuity argument shows that the formula is also valid in this case.

B. Runge's theorem

We continue to write $K^c = \mathbb{C} \backslash K$.

Theorem 2 (Runge 1885). *Suppose K is compact in $\mathbb{C}$ and f is analytic on K; further, let $\epsilon > 0$. Then there exists a rational function R with poles in K^c such that*

$$|f(z) - R(z)| < \epsilon \qquad (z \in K).$$

Proof. We use Theorem 1 and approximate the integral by Riemann sums. Since f is analytic on K, there exists an open set $U \supset K$ in which f is analytic. Suppose Γ is determined and consider

$$f(\zeta)/(\zeta - z) \quad \text{for } (\zeta, z) \in \Gamma \times K.$$

This function is continuous on $\Gamma \times K$, hence uniformly continuous. It follows that for each $\epsilon' > 0$ there exists a $\delta > 0$ such that

$$\left| \frac{f(\zeta)}{\zeta - z} - \frac{f(\zeta')}{\zeta' - z} \right| < \epsilon', \quad \text{whenever } z \in K \text{ and } \zeta, \zeta' \in \Gamma, |\zeta - \zeta'| < \delta.$$

We partition Γ into subintervals Γ_j of length $< \delta$ and choose $\zeta_j \in \Gamma_j$. Then

$$\left| \frac{1}{2\pi i} \int_{\Gamma_j} \frac{f(\zeta)}{\zeta - z}\, d\zeta - \frac{1}{2\pi i} \int_{\Gamma_j} \frac{f(\zeta_j)}{\zeta_j - z}\, d\zeta \right| < \frac{\epsilon'}{2\pi} |\Gamma_j| \qquad (z \in K),$$

and summing over j, we find that

$$|f(z) - R(z)| < \frac{\epsilon'}{2\pi} |\Gamma| \qquad (z \in K).$$

Here R is the rational function

$$R(z) = \frac{1}{2\pi i} \Sigma_j \frac{f(\zeta_j)}{\zeta_j - z} \int_{\Gamma_j} d\zeta,$$

which has simple poles at the points $\zeta_j \in \Gamma_j \subset K^c$. If we choose $\epsilon' = 2\pi\epsilon/|\Gamma|$, the assertion of Theorem 2 follows.

As mentioned above, R has the form $R(z) = \Sigma_j \dfrac{c_j}{z - \zeta_j}$, where $\zeta_j \in \Gamma$. If we choose a smaller ϵ, the ζ_j become more dense on Γ but do not move closer to K. The next section also is concerned with the location of the poles.

C. The pole shifting method

In this method, which also goes back to Runge [1885, p. 236], a rational function R_1 is approximated *on* K by another rational function R_2 whose poles lie elsewhere. This is *not always* possible. For example, if $K = \{z: |z| = 1\}$, $R_1(z) = 1/z$, and if R_2 is required to have a pole only at $z = 2$, then $\max_{z \in K} |R_1(z) - R_2(z)| < 1$ is not possible. One can see this immediately from

$$\int_K R_1(z)dz - \int_K R_2(z)dz = \int_K [R_1(z) - R_2(z)]\,dz,$$

since the left-hand side equals $2\pi i$, while the absolute value of the right-hand side would be less than 2π.

In view of the application in Chapter IV, §2, we now replace K by a general set M.

Theorem 3. *Suppose M is an arbitrary set in $\mathbb{C}$ and γ is a Jordan arc such that $\gamma \cap \overline{M} = \phi$. Let z_1 and z_2 denote the endpoints of γ. Then, for every polynomial P and $\epsilon > 0$, there exists a polynomial Q such that*

$$\left|P\left(\frac{1}{z - z_1}\right) - Q\left(\frac{1}{z - z_2}\right)\right| < \epsilon \qquad (z \in M). \tag{1.1}$$

Proof. Our assumption implies that dist $(\gamma, z) \geqslant \delta > 0$ for all $z \in M$. Suppose $z_1 = \zeta_0, \zeta_1, \zeta_2, \ldots, \zeta_N = z_2$ are points on γ such that $|\zeta_j - \zeta_{j+1}| < \delta$ ($j = 0, 1, \ldots, N-1$). Since $(z - z_1)^{-1}$ is analytic in $\{z: |z - \zeta_1| \geqslant \delta\} \cup \{\infty\}$, there exists, for $\eta > 0$, a polynomial p such that

$$\left|\frac{1}{z - z_1} - p\left(\frac{1}{z - \zeta_1}\right)\right| < \eta \quad \text{for } \{z: |z - \zeta_1| \geqslant \delta\}.$$

In particular, the last inequality holds for $z \in M$. Consequently we can find a polynomial P_1 such that

$$\left|P\left(\frac{1}{z - z_1}\right) - P_1\left(\frac{1}{z - \zeta_1}\right)\right| < \frac{\epsilon}{N} \quad \text{for } z \in M.$$

Analogously it follows that

$$\left|P_1\left(\frac{1}{z-\zeta_1}\right) - P_2\left(\frac{1}{z-\zeta_2}\right)\right| < \frac{\epsilon}{N} \quad \text{for } z \in M,$$

and finally

$$\left|P_{N-1}\left(\frac{1}{z-\zeta_{N-1}}\right) - P_N\left(\frac{1}{z-z_2}\right)\right| < \frac{\epsilon}{N} \quad \text{for } z \in M.$$

Thus (1.1) holds with $Q = P_N$.

Now we apply Theorem 3 with $M = K$ to Runge's theorem. There $R(z) = \Sigma_j\, c_j/(z - \zeta_j)$. The poles ζ_j lie on Γ, but according to Theorem 3, they can be shifted to points ζ_j^*, as long as ζ_j and ζ_j^* lie in the same component of K^c:

$$\left|\frac{c_j}{z-\zeta_j} - Q_j\left(\frac{1}{z-\zeta_j^*}\right)\right| < \epsilon \cdot 2^{-j} \quad (z \in K)$$

with certain polynomials Q_j. The rational function $R^*(z) = \Sigma_j Q_j\left(\dfrac{1}{z-\zeta_j^*}\right)$ then satisfies $|f(z) - R^*(z)| < 2\epsilon$ $(z \in K)$, and its poles are located at the points ζ_j^*.

Corollary 1 to Runge's theorem: *If one selects a point z_j in each component of K^c, then R can be chosen such that its poles occur at the points z_j.*

Clearly one component of K^c is unbounded. Suppose the pole corresponding to it occurs at $z_0 = m + 1$, where $m := \max\ \{|z|\colon z \in K\}$. Its contribution to R is of the form $P(1/(z - z_0))$, which is analytic for $|z| \leqslant m$. Hence there exists a polynomial Q such that

$$|P(1/(z - z_0)) - Q(z)| < \epsilon \quad \text{for } |z| \leqslant m$$

and thus for $z \in K$.

Corollary 2 to Runge's theorem: *For the unbounded component of K^c one can choose the point at ∞ as the location of the corresponding pole.*

In the special case where K^c consists of the unbounded component only, one obtains a result that we have proved earlier in Chapter II, §3, B (Runge's little theorem).

Theorem 4. *Suppose K is compact in $\mathbb{C}$ and K^c is connected; suppose further that f is analytic on K. Then, for each $\epsilon > 0$, there exists a polynomial P such that*

$$|f(z) - P(z)| < \epsilon \quad (z \in K).$$

Our goal is now to weaken the requirement in Theorems 2 and 4 that f be analytic on K and instead require only that f be continuous on K and analytic at the interior points of K. First we deal with polynomial approximation, before we continue to study (in §3) the approximation by rational functions.

§2. Mergelyan's theorem

This great theorem, proved by Mergelyan in 1951, completes a long chain of theorems about approximation by polynomials; the names Runge, Walsh, Keldysh, and Lavrentiev are connected with it. (Literature: Mergelyan [1951] and [1952, Chapter I], Rudin [1974, Chapter 20], Walsh [1969, Appendix 1]. These works use tools from classical analysis, whereas Carleson [1964] gives a functional-theoretic proof; Bishop [1960] as well as Glicksberg and Wermer [1963] should be cited as forerunners of it. We follow the classical route.)

A. Formulation of the result; special cases; consequences

Theorem 1 (Mergelyan 1951). *Suppose K is compact in $\mathbb{C}$ and K^c is connected; suppose further f is continuous on K and analytic in K°. Then, for $\epsilon > 0$, there exists a polynomial P such that*

$$|f(z) - P(z)| < \epsilon \qquad (z \in K). \tag{2.1}$$

Concerning the strength of the hypotheses, we can say that f must necessarily be continuous on K and analytic in K° in order for a sequence $\{P_n\}$ of polynomials to exist with $P_n(z) \rightrightarrows f(z)$ $(z \in K)$. The connectedness of K^c is also necessary, as we have already seen earlier; see Chapter II, §3, B.

The following *special cases* are included in Theorem 1: K is a closed interval $[a, b]$ (Weierstrass); $K = \{z: |z| \leqslant 1\}$ (Fejér polynomials, arithmetic means of the partial sums of the power series of f at 0); $K = \overline{G}$, where G is a Jordan domain and thus ∂G is a Jordan curve (Walsh 1926); $K = \Gamma$, where Γ is a Jordan arc (Walsh 1926). Incidentally, the last special case would be false in $\mathbb{C}^2$ (Rudin [1956]). Also included are the special cases $K^\circ = \phi$ (Lavrentiev 1934) and $K = \overline{K^\circ}$ (Keldysh 1945).

Further we mention three *consequences* of Theorem 1.

a) *If Γ is a Jordan curve,* $0 \in \operatorname{int} \Gamma$, *and f is continuous on Γ, then for $\epsilon > 0$ there exists a $P(z) = \Sigma_{n=-N}^{N} a_n z^n$ such that $|f(z) - P(z)| < \epsilon$ $(z \in \Gamma)$.*

For the proof we use a conformal mapping g of int Γ onto $\mathbb{D} = \{w: |w| < 1\}$; here we assume $g(0) = 0$. It has a continuous extension to $\overline{\text{int}\,\Gamma}$, just as $h := g^{-1}$ has a continuous extension to $\overline{\mathbb{D}}$. Since $f \circ h$ is continuous on $\partial\mathbb{D}$, there exists a trigonometric polynomial T such that $|f(h(w)) - T(\phi)| < \epsilon/2$ $(w = e^{i\phi})$. Now T can be written as a polynomial in w and $1/w$, so that

$$|f(h(w)) - \Sigma_{m=-M}^{M} A_m w^m| < \epsilon/2 \qquad (|w| = 1)$$

or

$$|f(z) - \Sigma_{m=-M}^{M} A_m (g(z))^m| < \epsilon/2 \qquad (z \in \Gamma).$$

According to Theorem 1, each function $(g(z))^m$ $(m \geqslant 0)$ can be approximated on Γ arbitrarily closely by polynomials, and for $(g(z))^m = (g(z)/z)^m \cdot z^m$ $(m < 0)$ this is true at least for the first factor. Our assertion a) follows from this.

b) *If Γ is a rectifiable Jordan curve, G =* int Γ, *and f is continuous on $\overline{G}$ and analytic in G, then the generalized Cauchy integral theorem holds:* $\int_\Gamma f(z)dz = 0$.

For each polynomial P we have that $\int_\Gamma f = \int_\Gamma (f - P) + \int_\Gamma P$. The second summand vanishes, and the first one can be made arbitrarily small by Theorem 1.

c) Using Theorem 1, one can prove the following generalization. *There exists a fixed power series that converges for $|z| < 1$ and whose partial sums $s_n(z)$ have the following property:*

For each compact set K with $K \cap \{z: |z| \leqslant 1\} = \phi$ and with connected K^c, and for each function f continuous on K and analytic in K°, there exists a sequence $\{n_k\}$ of indices such that

$$s_{n_k}(z) \Rightarrow f(z) \qquad (z \in K; k \to \infty).$$

The result is due to Chui and Parnes [1971]; Luh has extended it to matrix transformations [1974], [1976].

B. Preparations for the proof

We gather some auxiliary results. Later, Parts B_1 and B_2 will be used repeatedly.

B_1. Tietze's extension theorem

Suppose X is a locally compact Hausdorff space, K is a compact subset of X, and $f\colon K \mapsto \mathbb{C}$ is continuous on K. Then there exists a function $F\colon X \mapsto \mathbb{C}$, continuous on X and with compact support, such that $F(x) = f(x)$ for $x \in K$.

The extension F can be determined in such a way that

$$\max \{|F(x)| : x \in X\} = \max \{|f(x)| : x \in K\};$$

see, for example, Rudin [1974, p. 422]. For X we shall mostly use $\mathbb{C}$.

B_2. A representation formula

In the proof of Mergelyan's theorem nonanalytic functions occur, which are to be represented by an integral formula with the Cauchy kernel. Suppose now that G is an open set in $\mathbb{R}^2$ and $f\colon G \mapsto \mathbb{C}$ is continuously differentiable in G. We introduce two differential operators:

$$\partial f := \frac{1}{2}\left(\frac{\partial f}{\partial x} - i\,\frac{\partial f}{\partial y}\right), \qquad \bar{\partial} f := \frac{1}{2}\left(\frac{\partial f}{\partial x} + i\,\frac{\partial f}{\partial y}\right).$$

They also map G into $\mathbb{C}$. If we write $f = u + iv$, then

$$\bar{\partial} f = 0 \Leftrightarrow u_x = v_y,\ u_y = -v_x \Leftrightarrow f \text{ analytic in } G.$$

If this is the case, then $f'(z) = (\partial f)(z)$.

The following representation formula is sometimes called the Pompeiu formula.

Theorem 2. *Suppose the function $f\colon \mathbb{R}^2 \mapsto \mathbb{C}$ is continuously differentiable in $\mathbb{R}^2$ and has compact support. Then*

$$f(z) = -\frac{1}{\pi} \iint_{\mathbb{R}^2} \frac{(\bar{\partial} f)(\zeta)}{\zeta - z}\, dm_\zeta \tag{2.2}$$

for all $z \in \mathbb{R}^2$.

Proof. We substitute $\zeta = z + re^{i\phi}$ for the fixed $z \in \mathbb{R}^2$ and obtain $f(\zeta) = f(z + re^{i\phi}) = F(r, \phi)$. By the chain rule, we find that

$$(\bar{\partial} f)(\zeta) = \frac{1}{2}\, e^{i\phi}\left(F_r + \frac{i}{r}\, F_\phi\right).$$

Hence the right-hand side of (2.2) is the limit ($\epsilon \to 0$) of

$$-\frac{1}{2\pi}\int_{r=\epsilon}^{\infty}\int_{\phi=0}^{2\pi}(F_r + \frac{i}{r}F_\phi)\,d\phi\,dr.$$

Since F, as a function of ϕ, has period 2π, it follows that $\int F_\phi d\phi = 0$. Hence the right-hand side of (2.2) equals

$$\lim_{\epsilon\to 0}\frac{1}{2\pi}\int_{\phi=0}^{2\pi}F(\epsilon,\phi)d\phi = f(z),$$

because $F(\epsilon, \phi) \Rightarrow f(z) \quad (\epsilon \to 0)$.

B_3. Koebe's ¼-theorem

Let S denote the class of functions f that are univalent in $\mathbb{D} = \{z: |z| < 1\}$ and have an expansion of the form $f(z) = z + a_2 z^2 + \ldots$ at $z = 0$.

Koebe's theorem. *In $\mathbb{D}$ each $f \in S$ assumes each value ω with $|\omega| < 1/4$.*

In other words, the range of f in $\mathbb{D}$ covers the disk $\{\omega: |\omega| < ¼\}$. The number ¼ is best possible, as the function $f(z) = z + 2z^2 + 3z^3 + \ldots = z/(1 - z)^2$ shows. See, for example, Pommerenke [1975, p. 22].

We also require a corollary:

Corollary. *Suppose the function f is a conformal mapping of $\{z: |z| > 1\}$ onto a domain G with $\infty \in G$; suppose the expansion of f at ∞ is of the form*

$$f(z) = az + a_0 + a_1/z + \ldots .$$

Then the diameter of ∂G is $\leqslant 4|a|$.

The number 4 is sharp: For $f(z) = z + \frac{1}{z}$, we see that $\partial G = [-2, +2]$. For the *proof* we assume that f omits the values c and c' for z in $\{z: |z| > 1\}$. Consider

$$F(z) := \frac{a}{f(1/z) - c} \quad \text{for } z \in \mathbb{D};$$

clearly $F \in S$ and F does not assume the value $a/(c' - c)$ in $\mathbb{D}$, so that

$|a/(c' - c)| \geqslant 1/4$ by the ¼-theorem; that is, $|c' - c| \leqslant 4|a|$. The assertion follows from this.

Incidentally, the diameter of ∂G is always $\geqslant |a|$.

B_4. Mergelyan's lemma

This is a technical lemma whose significance will become clear only in the course of the proof. Suppose K is our compact set and

$$D := \{\zeta : |\zeta - \zeta_0| < r\}$$

is a disk. The *problem* is to approximate the Cauchy kernel $1/(z - \zeta)$ for $z \in K$ and $\zeta \in D$ by polynomials. In general, such an approximation is not possible, for example when $K = \{z : |z - \zeta_0| = 2r\}$; an additional topological condition is needed.

Mergelyan's lemma. *Suppose there exists a Jordan arc Γ from ζ_0 to ∞ such that $\Gamma \cap K = \phi$. Then there exist a polynomial p and a constant b with the following property. If one writes*

$$P_\zeta(z) := p(z) + (\zeta - b)(p(z))^2,$$

then

$$|P_\zeta(z)| < 100/r \qquad (z \in K; \zeta \in D), \tag{2.3}$$

$$|P_\zeta(z) - 1/(z - \zeta)| < 1000r^2/|z - \zeta|^3 \qquad (z \in K; \zeta \in D). \tag{2.4}$$

Note that the polynomial P_ζ in z and ζ depends on ζ in a particularly clear way. Note further that (2.4) follows from (2.3) if $z \in K, \zeta \in D$, and $|z - \zeta| < 2r$:

$$|z - \zeta|^3 \, |P_\zeta(z) - \frac{1}{z - \zeta}| < 8r^3 \cdot \frac{100}{r} + 4r^2 < 1000r^2.$$

For small $|z - \zeta|$ one uses (2.3), and for large $|z - \zeta|$ one uses (2.4).

The **proof** of the lemma is in two steps. *1st step:* First we define an analytic function q (not a polynomial) such that

$$Q_\zeta(z) := q(z) + (\zeta - b)(q(z))^2 \tag{2.5}$$

has properties analogous to (2.3) and (2.4). To this end let γ denote the part of Γ from ζ_0 to the first point of intersection of Γ with ∂D, and let $\gamma^c = \mathbb{C}\backslash\gamma$.

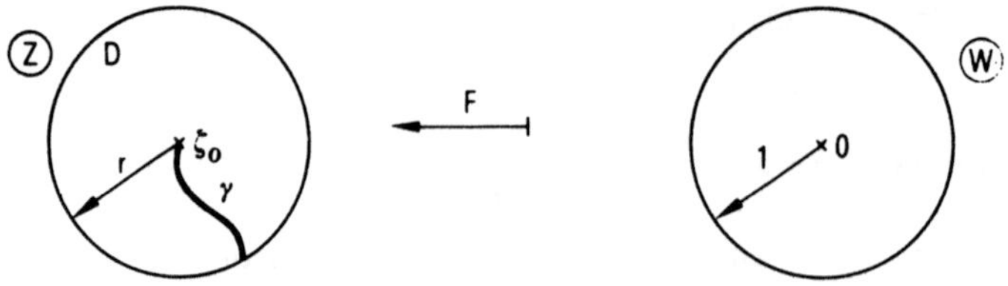

$$F(w) = \frac{a}{w} + \sum_{n=0}^{\infty} c_n w^n.$$

Suppose F is the normalized conformal mapping of $\{w\colon |w| < 1\}$ onto $\gamma^c \cup \{\infty\}$ such that $0 \mapsto \infty$ and $a > 0$. We write

(2.6) $$q(z) := \frac{F^{-1}(z)}{a} \quad (z \in \gamma^c) \quad \text{and} \quad b := \frac{1}{2\pi i}\int_{\partial D} zq(z)dz$$

and define $Q_\zeta(z)$ for $z \in \gamma^c$ and $\zeta \in D$ according to (2.5).

Then we have the following *properties:*

a) The corollary in B_3 yields $r \leqslant \operatorname{diam} \gamma \leqslant 4a$; hence $a \geqslant r/4$.

b) This implies first that

$$|q(z)| < 1/a \leqslant 4/r \qquad (z \in \gamma^c).$$

Further, q has the expansion $q(z) = 1/z + \ldots$ at ∞, so that $\int_{\partial \mathbb{D}} q(z)dz = 2\pi i$, and therefore

$$|b - \zeta_0| = \Big|\frac{1}{2\pi i}\int_{\partial D}(z - \zeta_0)q(z)dz\Big| \leqslant \frac{1}{2\pi}\, r \cdot \frac{4}{r} \cdot 2\pi r = 4r.$$

c) We now have

(2.3′) $$|Q_\zeta(z)| < \frac{4}{r} + 5r \cdot (4/r)^2 = 84/r \qquad (z \in \gamma^c; \zeta \in D)$$

as the analogous property to (2.3). In order to find the statement analogous to (2.4), we expand q into a Laurent series about the fixed point $\zeta \in D$:

$$q(z) = \frac{1}{z - \zeta} + \frac{A_2(\zeta)}{(z - \zeta)^2} + \ldots .$$

The first coefficient is 1, since $zq(z) \to 1$ $(z \to \infty)$, and the series converges in any case for $|z - \zeta| > 2r$. Moreover,

$$A_2(\zeta) = \frac{1}{2\pi i} \int_k (z - \zeta)q(z)dz = b - \zeta \cdot 1,$$

where the integration is over a large circle k. For the fixed $\zeta \in D$ and $z \in \gamma^c$, we now consider the auxiliary function

$$H(z) := \left[Q_\zeta(z) - \frac{1}{z - \zeta}\right] (z - \zeta)^3$$

$$= \left[q(z) + (\zeta - b)(q(z))^2 - \frac{1}{z - \zeta}\right] (z - \zeta)^3$$

$$= \left[\frac{b - \zeta}{(z - \zeta)^2} + \frac{\zeta - b}{(z - \zeta)^2} + \text{decreasing powers}\right] (z - \zeta)^3.$$

It follows that H is analytic at ∞, so the maximum principle can be applied:

$$|H(z)| \leqslant \max \{|H(z)|: z \in \gamma\} \leqslant \frac{84}{r} (2r)^3 + (2r)^2 = 676\, r^2 \qquad (z \in \gamma^c).$$

Thus the statement analogous to (2.4) holds:

(2.4′) $$|Q_\zeta(z) - \frac{1}{z - \zeta}| < \frac{700 r^2}{|z - \zeta|^3} \quad (z \in \gamma^c; \zeta \in D).$$

2nd *step*. In order to arrive at (2.3) and (2.4), we approximate q by a polynomial. Since q is analytic in γ^c, it is also analytic inside and on each level curve Γ_ρ of Γ.

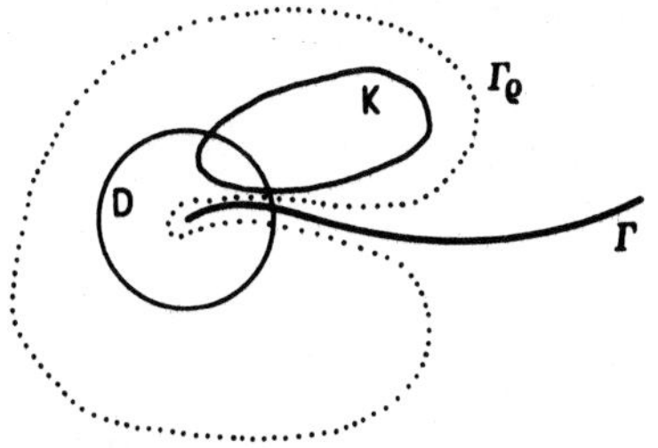

If ρ is sufficiently close to 1, then Γ_ρ encloses the compact set K, which is disjoint from Γ. According to Runge's little theorem, there exists for $\epsilon > 0$ a polynomial p such that

$$|q(z) - p(z)| < \epsilon, \quad |(q(z))^2 - (p(z))^2| < \epsilon \quad (z \in K).$$

If we form $P_\zeta(z) = p(z) + (\zeta - b)(p(z))^2$ with this p, then

$$|P_\zeta(z) - Q_\zeta(z)| \leqslant |p(z) - q(z)| + |\zeta - b|\, |(p(z))^2 - (q(z))^2| < \epsilon + 5r\epsilon$$

$$(z \in K; \zeta \in D).$$

Now (2.3′) and (2.4′) imply

$$|P_\zeta(z)| < \frac{84}{r} + \epsilon(1 + 5r) < \frac{100}{r} \quad (z \in K; \zeta \in D)$$

whenever $\epsilon < \epsilon_1$, and

$$\left| P_\zeta(z) - \frac{1}{z - \zeta} \right| < \frac{700r^2}{|z - \zeta|^3} + \epsilon(1 + 5r) < \frac{1000r^2}{|z - \zeta|^3} \quad (z \in K; \zeta \in D)$$

whenever $\epsilon < \epsilon_2$. Note that $|z - \zeta|$ has a fixed upper bound for these values of z, ζ.

Consequently (2.3) and (2.4) are valid for the polynomial p chosen for a suitable $\epsilon > 0$.

C. Proof of Mergelyan's theorem

According to Tietze's theorem, f can be extended to a function that is continuous in $\mathbb{C}$ and has compact support. We assume such an extension has already taken place; the resulting function is again denoted by f. As usual, we let

$$\omega(\delta) := \sup \{|f(z_1) - f(z_2)| : |z_1 - z_2| \leqslant \delta\} \quad (\delta > 0).$$

Since f is uniformly continuous in $\mathbb{C}$, it follows that $\omega(\delta) \to 0$ $(\delta \to 0)$.

We show: For each prescribed $\delta > 0$ there exists a polynomial P such that

$$|f(z) - P(z)| < 10\,000\; \omega(\delta) \quad (z \in K);$$

this establishes the theorem. We carry out the proof in two steps:

(1): Construction of a nonanalytic function Φ that is close to f on $\mathbb{C}$;
(2): Construction of a polynomial P that is close to Φ on K.

1st step (application of a smoothing operation). First we define

$$G := \{z \in K\colon \operatorname{dist}(z, \partial K) > \delta\}.$$

This is an open set containing all points of K° that are "far inside" K; it is possible that G is empty.

Next we construct a function Φ that is continuously differentiable in $\mathbb{C}$, has compact support, and possesses the following properties:

$$(2.7)\qquad \begin{array}{lll} \text{(a)} & |f(z) - \Phi(z)| \leqslant \omega(\delta) & (z \in \mathbb{C}), \\ \text{(b)} & |(\bar\partial\Phi)(z)| \leqslant 2\omega(\delta)/\delta & (z \in \mathbb{C}), \\ \text{(c)} & \Phi(z) = f(z) & (z \in G). \end{array}$$

To this end we start with

$$a(r) = \begin{cases} \dfrac{3}{\pi\delta^2}\left(1 - \dfrac{r^2}{\delta^2}\right)^2 & 0 \leqslant r \leqslant \delta, \\ 0 & r \geqslant \delta, \end{cases}$$

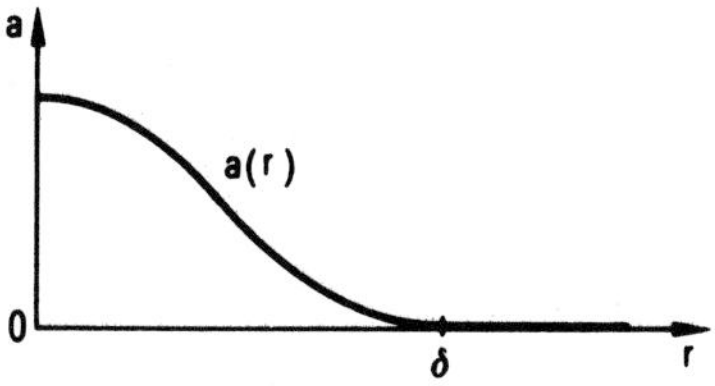

write $A(z) = a(|z|)$ $(z \in \mathbb{C})$, and note that

$$(2.8)\qquad \iint_{\mathbb{R}^2} A = 1, \quad \iint_{\mathbb{R}^2} \bar\partial A = 0, \quad \iint_{\mathbb{R}^2} |\bar\partial A| = \frac{24}{15\delta} < \frac{2}{\delta}.$$

The first and last relation can be verified by introducing polar coordinates; and $\iint_{\mathbb{R}^2} (A_x + iA_y)dx\,dy$ vanishes, because A has compact support. With this function A we define

$$(2.9)\qquad \Phi(z) := \iint_{\mathbb{R}^2} f(z - \zeta)A(\zeta)dm_\zeta = \iint_{\mathbb{R}^2} f(\zeta)A(z - \zeta)dm_\zeta \qquad (z \in \mathbb{C}),$$

a convolution of f with the smoothing function A. Note that because $A(z - \zeta) = 0$ for $|z - \zeta| \geqslant \delta$, we have

$$\Phi(z) = \iint_{\{\zeta:\, |\zeta - z| < \delta\}} f(\zeta)A(z - \zeta)dm_\zeta \quad (z \in \mathbb{C}).$$

Consequently $\Phi(z) = 0$, if the distance from z to the support of f is $\geqslant \delta$. In particular Φ, like f, has compact support.

Now we prove (2.7). Concerning (a): We have

$$\Phi(z) - f(z) = \iint_{\mathbf{R}^2} [f(z-\zeta) - f(z)]\, A(\zeta) dm_\zeta$$

$$= \iint_{\{\zeta:\ |\zeta| \leqslant \delta\}} [f(z-\zeta) - f(z)]\, A(\zeta) dm_\zeta .$$

Now (2.8) implies (a).

The derivatives Φ_x and Φ_y necessary for (b) can be obtained by differentiating under the integral (2.9). For $h > 0$ we have, for example,

$$\frac{\Phi(z+h) - \Phi(z)}{h} = \iint_{\mathbf{R}^2} f(\zeta) \frac{A(z+h-\zeta) - A(z-\zeta)}{h} dm_\zeta .$$

For $z, \zeta \in \mathbf{C}$ and $h > 0$ the integrand has a fixed upper bound, and the limit $(h \to 0)$ is $f(\zeta) A_x(z-\zeta)$. Since we integrate over a compact set, the Lebesgue Bounded Convergence Theorem applies:

$$(\bar{\partial}\Phi)(z) = \iint_{\mathbf{R}^2} f(\zeta)(\bar{\partial}A)(z-\zeta) dm_\zeta = \iint_{\mathbf{R}^2} f(z-\zeta)(\bar{\partial}A)(\zeta) dm_\zeta$$

$$= \iint_{\mathbf{R}^2} [f(z-\zeta) - f(z)]\, (\bar{\partial}A)(\zeta) dm_\zeta ,$$

by (2.8). The third relation in (2.8) now yields (2.7) (b).

In order to establish (2.7) (c), let $z \in G$, so that f is analytic in the closed δ-neighborhood of z. The mean-value property implies

$$\int_{\phi=0}^{2\pi} f(z - \rho e^{i\phi}) d\phi = 2\pi f(z) \qquad (\rho \leqslant \delta).$$

Hence

$$\Phi(z) = \iint\limits_{\{\zeta:\ |\zeta - z| < \delta\}} f(\zeta) A(z - \zeta) dm_\zeta$$

$$= \int\limits_{\rho=0}^{\delta} \int\limits_{\phi=0}^{2\pi} f(z - \rho e^{i\phi}) a(\rho) \rho \, d\rho \, d\phi = 2\pi f(z) \int\limits_{\rho=0}^{\delta} a(\rho) \rho \, d\rho$$

$$= f(z) \iint\limits_{\mathbb{R}^2} A(\zeta) dm_\zeta = f(z),$$

by (2.8). This proves (2.7) completely.

Before we begin the second step, we draw a conclusion from (2.7) (c). We apply the integral formula (2.2) to Φ and obtain

$$\Phi(z) = -\frac{1}{\pi} \iint\limits_{\mathbb{R}^2} \frac{(\bar{\partial}\Phi)(\zeta)}{\zeta - z} \, dm_\zeta.$$

On the open set $G \subset K^\circ$ the function $\Phi = f$ is analytic, hence $\bar{\partial}\Phi = 0$, and integration over G is not necessary:

The function Φ with the properties (2.7) has the representation

(2.10) $$\Phi(z) = -\frac{1}{\pi} \iint\limits_{X} \frac{(\bar{\partial}\Phi)(\zeta)}{\zeta - z} \, dm_\zeta \qquad (z \in \mathbb{C}),$$

where

$$X := \{\text{support of } \Phi\} \cap G^c$$

is a compact set.

2nd *step.* Finally, in order to approximate Φ by a polynomial P, we approximate the kernel $1/(\zeta - z)$ in (2.10) by a function $R_\zeta(z)$, which is piecewise a polynomial.

First we partition X into finitely many disjoint subsets X_j such that $\cup X_j = X$, as follows. For $\zeta \in X$ we have $\zeta \in G^c$, hence $\text{dist}(\zeta, K^c) \leqslant \delta$. This implies that the compact set X can be covered by finitely many open disks $D_1, \ldots, D_N$ with radius 2δ, whose centers M_j lie in K^c. By assumption K^c is connected, so that there exist Jordan arcs Γ_j that connect M_j with ∞ without meeting K.

Now we write

$$X_1 := X \cap D_1, X_j := X \cap D_j \setminus (X_1 \cup X_2 \cup \ldots \cup X_{j-1}) \qquad (j = 2, \ldots, N);$$

these sets $X_j \subset D_j$ are disjoint, and $\cup X_j = X$.

Suppose the construction in Mergelyan's lemma has been carried out for each disk D_j $(r = 2\delta)$; this yields polynomials $P_{\zeta,j}$ with the properties (2.3) and (2.4). The function R_ζ mentioned above is now pieced together from the $P_{\zeta,j}$:

$$R_\zeta(z) := P_{\zeta,j}(z) \quad \text{for} \quad \zeta \in X_j;$$

this defines $R_\zeta(z)$ for $\zeta \in X$ and $z \in \mathbb{C}$. The function R_ζ satisfies

$$\text{(2.11)} \qquad |R_\zeta(z)| < 50/\delta \qquad (z \in K; \zeta \in X),$$

$$\text{(2.12)} \qquad |R_\zeta(z) - \frac{1}{z - \zeta}| < \frac{4000\delta^2}{|z - \zeta|^3} \qquad (z \in K; \zeta \in X),$$

as can be seen from (2.3) and (2.4).

Finally, we define the desired polynomial P by

$$P(z) := \frac{1}{\pi} \iint_X (\bar{\partial}\Phi)(\zeta) R_\zeta(z) dm_\zeta = \frac{1}{\pi} \Sigma_j \iint_{X_j} (\bar{\partial}\Phi)(\zeta) P_{\zeta,j}(z) dm_\zeta .$$

Here each summand is a polynomial in z, hence P is a polynomial.

It remains to estimate $|P(z) - \Phi(z)|$ for $z \in K$. Using (2.10) and (2.7) (b), we obtain

$$|P(z) - \Phi(z)| = \left| \frac{1}{\pi} \iint_X \left[R_\zeta(z) - \frac{1}{z - \zeta}\right] (\bar{\partial}\Phi)(\zeta) dm_\zeta \right|$$

$$\leqslant \frac{2\omega(\delta)}{\pi\delta} \iint_X \left| R_\zeta(z) - \frac{1}{z - \zeta} \right| dm_\zeta = \frac{2\omega(\delta)}{\pi\delta} \left[\iint_{X^{(1)}} + \iint_{X^{(2)}} \right],$$

where, for fixed $z \in K$,

$$X^{(1)} := \{\zeta \in X : |\zeta - z| < 4\delta\},$$

$$X^{(2)} := \{\zeta \in X : |\zeta - z| \geqslant 4\delta\}.$$

We estimate the integral over $X^{(1)}$ by (2.11):

$$|\iint_{X^{(1)}}| \leqslant \frac{50}{\delta} \cdot \pi(4\delta)^2 + 2\pi \cdot (4\delta) = 808\,\pi\delta,$$

and we estimate the integral over $X^{(2)}$ by (2.12):

$$|\iint_{X^{(2)}}| < 2\pi \int_{\rho=4\delta}^{\infty} \frac{4000\delta^2}{\rho^2}\, d\rho = 2000\,\pi\delta.$$

Altogether we have

$$|P(z) - \Phi(z)| < \frac{2\omega(\delta)}{\pi\delta} \cdot 2808\,\pi\delta < 6000\,\omega(\delta) \qquad (z \in K).$$

If one combines this with (2.7) (a), the assertion at the beginning of Section C follows.

Remark about §2

Only recently Arakeljan and Martirosjan [1977] have investigated under which assumptions about the compact set K and the index set E each function f that is continuous on K and analytic in K° can be approximated arbitrarily closely by *lacunary polynomials* $P(z) = \Sigma_{k \in E} a_k z^k$. Here it is assumed that E contains the index zero and has density 1, also that K^c is connected. In case $0 \in K^\circ$, it is necessary that $E = \mathbb{N} \cup \{0\}$; but if $0 \notin K$ or if $0 \in \partial K$ (with additional conditions), it suffices that E has density 1.

For this area see also the recent results by Korevaar and Dixon [1977], [1978] on lacunary approximation.

§3. Approximation by rational functions

We take up anew the topic of rational approximation begun in §1, though under the weaker assumption that f is continuous on the compact set K and analytic in K°. The counterpart to Mergelyan's theorem is Vitushkin's theorem (1966); we can only report about it here. But we shall treat in detail several instructive examples, Bishop's localization theorem, and some of its applications.

To simplify the notation we introduce four function spaces:

$C(K) = \{f: f \text{ continuous on } K\}$ with norm $\|f\| = \max\,\{|f(z)|: z \in K\}$,

$A(K) = \{f \in C(K): f \text{ analytic in } K^\circ\}$,

$P(K) = \{f \in A(K)$: for each $\epsilon > 0$ there exists a polynomial P such that $\|f - P\| < \epsilon\}$,

$R(K) = \{f \in A(K)$: for each $\epsilon > 0$ there exists a rational function R such that $\|f - R\| < \epsilon\}$.

The poles of R are then automatically outside K. All spaces are algebras with the "uniform norm" $\|f\|$. This point of view is prevalent in Gamelin [1969]. Obviously

$$P(K) \subset R(K) \subset A(K) \subset C(K),$$

and by Mergelyan's theorem, $P(K) = A(K)$ if and only if K^c is connected. Analogously we now ask: When is $R(K) = A(K)$, or, more modestly: When does $f \in R(K)$ follow from $f \in A(K)$?

A. Swiss cheese

By this we understand compact sets K with infinitely many holes, so that K^c consists of infinitely many components. For these sets (and only for these, see Section C_2) it is possible that $R(K) \neq A(K)$.

A_1. Alice Roth's construction

The Swiss mathematician Alice Roth [1938, p. 96] constructed the first Swiss cheese as follows. Suppose $\mathbb{D} = \{z: |z| < 1\}$, and let $\Delta_j = \{z: |z - a_j| < r_j\}$, where $\overline{\Delta}_j \subset \mathbb{D}$, denote countably many open disks with the following properties:

a) The $\overline{\Delta}_j$ are pairwise disjoint;

b) $\Sigma r_j < 1$;

c) $\overline{\mathbb{D}} \setminus \cup \Delta_j$ contains no disk.

Swiss cheese

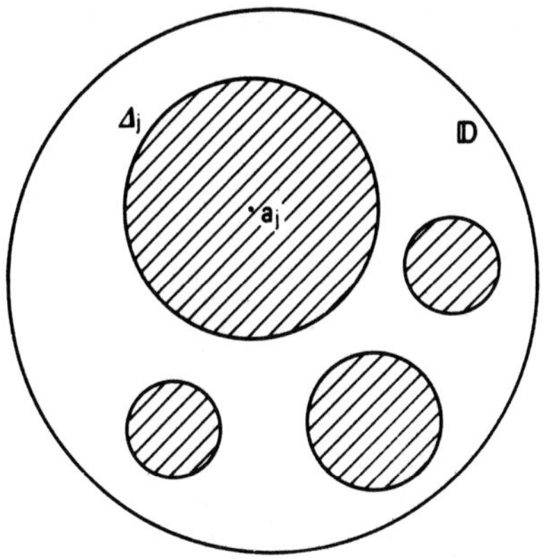

Such disks Δ_j can be chosen in many different ways. The remaining set

(3.1) $$K = \overline{\mathbb{D}} \setminus \cup_j \Delta_j$$

is Roth's Swiss cheese. Clearly K is compact, K contains, for example, all circles $\partial\Delta_j$ as well as $\partial\mathbb{D}$, $K^\circ = \phi$, and

$$K^c = (\cup\Delta_j) \cup \{z: |z| > 1\}$$

consists of countably many components. Incidentally, the two-dimensional Lebesgue measure of K can be arbitrarily close to π, if the r_j are chosen sufficiently small.

If the rational function R belongs to $R(K)$ and if $\partial\mathbb{D}$ and $\partial\Delta_j$ have positive orientation, then

$$\int_{\partial\mathbb{D}} R(z)\,dz = \Sigma_j \int_{\partial\Delta_j} R(z)\,dz;$$

on the right-hand side only finitely many summands do not vanish. Therefore, if $f \in R(K)$, then

$$\left|\int_{\partial\mathbb{D}} f(z)dz - \Sigma_j \int_{\partial\Delta_j} f(z)dz\right| = \left|\int_{\partial\mathbb{D}} [f(z) - R(z)]\,dz - \Sigma_j \int_{\partial\Delta_j} [f(z) - R(z)]\,dz\right|$$
$$\leqslant \|f - R\|\,(2\pi + \Sigma\, 2\pi r_j)$$

can be made arbitrarily small, so that

(3.2) $$\int_{\partial\mathbb{D}} f(z)dz = \Sigma_j \int_{\partial\Delta_j} f(z)dz \quad \text{for all} \quad f \in R(K).$$

Now we assume that, say, $0 \notin K$ and consider

$$f(z) = |z|/z = e^{-i\phi} \quad \text{for} \quad z = re^{i\phi} \in K.$$

Then $f \in C(K) = A(K)$, because $K^\circ = \phi$. Further, the left-hand side of (3.2) equals $2\pi i$, while the right-hand side has absolute value $\leqslant \Sigma\, 2\pi r_j < 2\pi$. Consequently $f \notin R(K)$:

For the compact set (3.1) *we have* $R(K) \neq A(K)$.

By varying Roth's idea, one can construct further compact sets, which have additional properties and allow further conclusions.

A_2. Swiss cheese with interior points

First suppose Γ is a Jordan arc with positive two-dimensional Lebesgue measure $\mu(\Gamma)$, and assume Γ lies inside $\mathbb{D}$ with the exception of one endpoint ($z = 1$). Suppose the Δ_j, where $\overline{\Delta}_j \subset \mathbb{D}$, are countably many open disks with radii r_j and with the properties:

a) The $\overline{\Delta}_j$ are pairwise disjoint;

b) $\Sigma\, r_j < \infty$;

c) $\overline{\Delta}_j \cap \Gamma$ is a point P_j;

d) Each point of Γ is an accumulation point of $\{\Delta_j\}$.

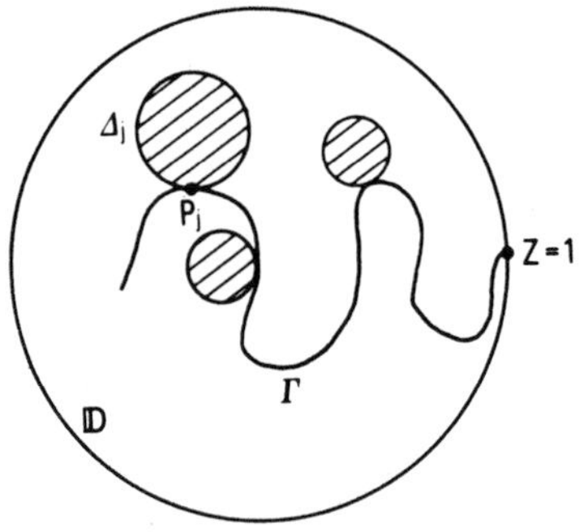

As in Roth's example, we set

$$K := \overline{\mathbb{D}} \setminus \cup_j \Delta_j.$$

The set K is compact and K^c has infinitely many components, but K° is not empty: Here K° is a simply connected domain whose closure is K.

As in Section A_1 one shows now that for each $f \in R(K)$ the equation

$$\int_{\partial \mathbb{D}} f(z)dz = \Sigma_j \int_{\partial \Delta_j} f(z)dz \tag{3.2}$$

must necessarily hold. But because $K^\circ \neq \phi$, we must consider a different function f, such as

$$f(z) = \iint_{\Gamma} \frac{1}{\zeta - z}\, dm_\zeta \qquad (z \in \mathbb{C}).$$

This function is continuous in $\mathbb{C}$ and even analytic outside Γ, so that certainly $f \in A(K)$; note that $\Gamma \subset \partial K$, by d). Integration over $\gamma = \{z: |z| = 2\}$ yields

$$\int_\gamma f(z)dz = \iint_\Gamma \left(\int_\gamma \frac{dz}{\zeta - z} \right) dm_\zeta = -2\pi i \cdot \mu(\Gamma) \neq 0;$$

consequently also $\int_{\partial \mathbf{D}} f(z)dz \neq 0$, whereas all integrals $\int_{\partial \Delta_j} f(z)dz$ vanish. This violates (3.2), and $f \notin R(K)$.

It follows again that $R(K) \neq A(K)$, even though $K^\circ \neq \phi$ and K° was topologically very simple.

A_3. Swiss cheese with two components

We modify the example in Section A_2 in such a way that both endpoints of the Jordon arc Γ lie on $\partial \mathbf{D}$, say at ± 1; the "stitched disk" results. Here again $R(K) \neq A(K)$. But K° now divides into *two* simply connected domains: $K^\circ = G_1 \cup G_2$. To judge rational approximation on the compact sets $K_1 = \overline{G}_1$, $K_2 = \overline{G}_2$ we use the following criterion.

Criterion. *If a compact set K has the property that each point* $z \in \partial K$ *is a boundary point of a component of* K^c, *then* $R(K) = A(K)$.

This is a consequence of Vitushkin's theorem (Gamelin [1969, p. 219]; Zalcman [1968, p. 108]); for our situation it leads to the observation:

In this case $R(K_j) = A(K_j)$ $(j = 1, 2)$, *but* $R(K_1 \cup K_2) \neq A(K_1 \cup K_2)$.

A_4. Accumulation of holes along the diameter of $\mathbf{D}$

In this example we line up the domains Δ_j along the diameter $[-1, +1]$; suppose the boundaries γ_j of Δ_j are Jordan curves. We assume:

a) The $\overline{\Delta}_j$ are pairwise disjoint;

b) diam $\gamma_j \to 0$ $(j \to \infty)$;

c) $\overline{\Delta}_j \cap [-1, +1]$ is a point P_j.

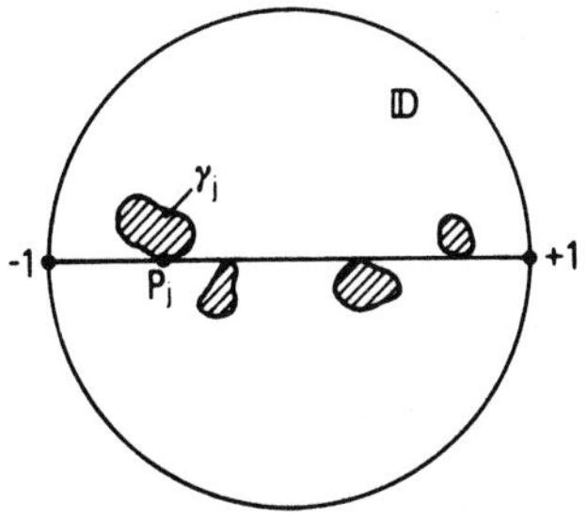

Again we let

$$K := \overline{\mathbf{D}} \setminus \cup_j \Delta_j$$

be our compact set. In this case one can show that for each function $f \in A(K)$ and each $\epsilon > 0$ there exists a function g_ϵ such that $\|f - g_\epsilon\|_K < \epsilon$, where g_ϵ

also belongs to $A(K)$ and is, in addition, analytic on $[-1, +1]$ (Gamelin [1969, p. 235]). The corollary to Theorem 4 (see Part C) can be applied to this function; hence there exists a rational function R such that $\|g_\epsilon - R\|_{K'} < \epsilon$, where $K' \supset K$.

Hence we have for each such compact set K that $R(K) = A(K)$.

Now suppose h is a homeomorphism from $\mathbb{C}$ to $\mathbb{C}$ such that $h(z) = z$ for $|z| \geqslant 1$ and such that the arc Γ from the example in A_3 is mapped onto $[-1, +1]$. Then the compact set in A_3 is tranformed into a compact set K of the form above. Considering $g = h^{-1}$, we see:

There are cases where $R(K) = A(K)$, but $R(g(K)) \neq A(g(K))$.

The property $R(K) = A(K)$ is not invariant under topological mappings of the plane.

All compact sets in Section A have the property that K^c consists of infinitely many components. Our next goal is the theorem that $R(K) = A(K)$ if K^c has only finitely many components. This will be accomplished with the help of the important localization theorem by Bishop, preparations for which will be found in the next section.

B. Preparations for Bishop's theorem

B_1. An integral transform

In the proof of Bishop's theorem, as was the case with Mergelyan's theorem, nonanalytic functions occur, which are represented by an integral with Cauchy kernel.

Theorem 1. *Suppose f is continuous in $\mathbb{R}^2$ and q is continuously differentiable in $\mathbb{R}^2$ and with compact support t_g; let*

(3.3) $$f_g(z) := \frac{1}{\pi} \iint_{\mathbb{R}^2} \frac{f(\zeta) - f(z)}{\zeta - z} (\bar{\partial} g)(\zeta) dm_\zeta \qquad (z \in \mathbb{R}^2).$$

Then

a) f_g *admits the representation*

(3.4) $$f_g(z) := f(z)g(z) + \frac{1}{\pi} \iint_{\mathbb{R}^2} \frac{f(\zeta)}{\zeta - z} (\bar{\partial} g)(\zeta) dm_\zeta \qquad (z \in \mathbb{R}^2);$$

b) f_g *is continuous in* $\mathbb{R}^2$, *and* $f_g(z) \to 0$ *for* $z \to \infty$;

c) f_g *is analytic in* t_g^c,

d) *if f is analytic in G, then so is f_g;*

e) *if the support t_g of g has diameter $\leqslant \delta$, then*

$$\|f_g\|_{\mathbf{R}^2} \leqslant 2\delta\ \omega_f(\delta)\ \|\bar{\partial} g\|_{\mathbf{R}^2}.$$

Proof. a) This follows from the representation formula (2.2) in § 2, applied to g.

b) If we write $\zeta = z - u$ in (3.4), then the second term becomes $\iint F(z-u)u^{-1}dm_u$, where F is continuous in $\mathbf{R}^2$ and has compact support. From this the continuity of f_g in $\mathbf{R}^2$ follows; (3.4) implies that $f_g(z) \to 0$ as $z \to \infty$.

c) For $z \notin t_g$ we have $f_g(z) = \dfrac{1}{\pi}\displaystyle\iint \frac{f(\zeta)}{\zeta - z}\,(\bar{\partial} g)(\zeta)dm_\zeta$, where the integral is over t_g. It follows that f_g is analytic in t_g^c. Because of b), f_g is even analytic at ∞.

d) For this part one forms $[f_g(z+h) - f_g(z)]/h$ and evaluates the limit as $h \to 0$ ($h \in \mathbf{C}$; $z \in G$). Lebesgue's theorem needs to be applied.

e) Considering c), b), and the maximum principle, one sees that $\|f_g\|_{\mathbf{R}^2}$ is assumed for some $z_0 \in t_g$. But then (3.3) implies

$$|f_g(z_0)| \leqslant \frac{1}{\pi}\ \omega_f(\delta)\|\bar{\partial} g\|_{\mathbf{R}^2} \iint_{t_g} \frac{dm_\zeta}{|\zeta - z_0|},$$

and the integral ($\zeta = z_0 + re^{i\phi}$) is bounded in magnitude by $\displaystyle\int_{\phi=0}^{2\pi}\int_{r=0}^{\delta} dr\, d\phi = 2\pi\delta$.

B_2. Partition of unity

This device also plays an important role in local theorems in topology and functional analysis. We proceed in two steps.

Lemma 1. *Suppose K is a compact set in $\mathbf{C}$ and $U \supset K$ is open. Then there exists a function $H \in C^\infty(\mathbf{R}^2)$ such that $0 \leqslant H(z) \leqslant 1$ ($z \in \mathbf{R}^2$) and*

$$H(z) = 1 \quad (z \in K), \qquad H(z) = 0 \quad (z \notin U).$$

Proof. For each $\zeta \in K$ we choose disks u_ζ, U_ζ such that $u_\zeta \subset U_\zeta \subset U$, and we choose functions $h_\zeta \in C^\infty(\mathbf{R}^2)$ *such that* $0 \leqslant h_\zeta(z) \leqslant 1$ ($z \in \mathbf{R}^2$) and

$$h_\zeta(z) = 1 \quad (z \in u_\zeta), \qquad h_\zeta(z) = 0 \quad (z \notin U_\zeta).$$

Suppose the disks u_{ζ_j} ($j = 1, 2, \dots, m$) cover K. Then the function

$$H(z) := 1 - \Pi_{j=1}^{m} (1 - h_{\zeta_j}(z)) \qquad (z \in \mathbf{R}^2)$$

has the desired properties.

We proceed with the partition of unity; we need it in the following form.

Theorem 2. *Suppose we have finitely many pairs of disks u_j, U_j $(u_j \subset U_j)$ about points ζ_j. Then there exist functions $\phi_j \in C^\infty(\mathbf{R}^2)$ such that $0 \leqslant \phi_j(z) \leqslant 1$ $(z \in \mathbf{R}^2)$ and*

$$\phi_j(z) = 0 \quad (z \notin U_j) \quad \textit{and} \quad \Sigma_j \phi_j(z) = 1 \quad (z \in \cup u_j).$$

On $\cup u_j$, the function that is identically equal to one is additively decomposed into the functions ϕ_j.

Proof. First we construct the h_{ζ_j} from above, and then we determine disks $u_j' \supset \bar{u}_j$ such that $h_{\zeta_j}(z) > 1/2$ for $z \in u_j'$. Then we use

$$K = \cup \bar{u}_j \quad \text{and} \quad U = \cup u_j'$$

in Lemma 1 and determine the function H there.

Finally, we let

$$\phi_j(z) = H(z)[\Sigma_j h_{\zeta_j}(z)]^{-1} h_{\zeta_j}(z) \qquad (z \in \mathbf{R}^2).$$

Each ϕ_j belongs to $C^\infty(\mathbf{R}^2)$, since in a neighborhood of $\bar{U}$ the sum Σ_j is greater than $1/4$ and since $H(z) = 0$ for $z \notin U$. The other properties of ϕ_j mentioned in Theorem 2 follow immediately; for example, $\Sigma_j \phi_j(z) = H(z) = 1$ if $z \in \cup u_j \subset K$.

C. Bishop's localization theorem and applications

The examples in Section A made it clear that the characterization of sets K for which $R(K) = A(K)$ would be difficult. Nevertheless one can show without much effort that the property $R(K) = A(K)$ is a "local property" of the compact set K.

C_1. The localization theorem

Theorem 3 (Bishop). *Suppose K is a compact set in $\mathbf{C}$ and f is continuous in $\mathbf{C}$. Suppose each $z \in K$ has a neighborhood U_z such that $f|_{K_z} \in R(K_z)$ for the compact set $K_z := K \cap \bar{U}_z$. Then $f \in R(K)$.*

Note that even though initially f is only required to be continuous on K (or on $\mathbb{C}$), the additional assumption that $f|_{K_z} \in R(K_z)$ for each $z \in K$ implies that $f \in A(K)$. There are several proofs of Theorem 3. The following proof is due to Garnett, and in §4 we shall give another proof that uses Roth's lemma.

Proof. *1st step.* First we partition f additively on K. For $z \in K$ we determine the neighborhood U_z according to the hypothesis of the theorem; obviously U_z can be chosen as a disk with center at z. Suppose u_z is the disk about z with half the radius: $u_z \subset U_z$. Finitely many of these disks cover K; we denote them by u_j $(j = 1, 2, \ldots, n)$ and construct the functions ϕ_j corresponding to u_j and U_j, according to Theorem 2. These functions satisfy

$$\Sigma_j \phi_j(z) = 1 \quad \text{for} \quad z \in V := \cup \bar{u}_j, \quad \text{where } V^\circ \supset K,$$

and $\phi_j(z) = 0$ for $z \notin U_j$. Finally, we let

$$C := \max_j \max_{z \in \mathbb{C}} \frac{1}{\pi} \iint_{\mathbb{R}^2} \frac{|(\bar{\partial}\phi_j)(\zeta)|}{|\zeta - z|} \, dm_\zeta < \infty.$$

In order to get the desired partition of f, we write

$$f_j(z) := f(z)\phi_j(z) + \frac{1}{\pi} \iint_{\mathbb{R}^2} \frac{f(\zeta)}{\zeta - z} (\bar{\partial}\phi_j)(\zeta) dm_\zeta \qquad (z \in \mathbb{R}^2), \tag{3.5}$$

so that $f_j = f_{\phi_j}$ in the notation of (3.4). For $z \in V$ we find that $\Sigma_j f_j(z) = f(z) + \Phi(z)$, where

$$\Phi(z) = \frac{1}{\pi} \iint_{\mathbb{R}^2} \frac{f(\zeta)}{\zeta - z} (\bar{\partial}\phi)(\zeta) dm_\zeta = \frac{1}{\pi} \iint_{\mathbb{R}^2 \setminus V} \frac{f(\zeta)}{\zeta - z} (\bar{\partial}\phi)(\zeta) dm_\zeta,$$

because $\phi = \Sigma\phi_j = 1$ on V. The function Φ certainly is analytic on V°, hence also on K. Thus we can apply Runge's theorem to Φ and obtain: $\Phi \in R(K)$. Therefore it is sufficient to show that $f_j \in R(K)$ $(j = 1, 2, \ldots, n)$.

2nd step. Next we approximate the functions f_j. To this end, f_j will be approximated by a function g_j, which is generated like f_j. First we refer back to the local approximating functions: We write $K_j = K \cap \bar{U}_j$, and for a given $\epsilon > 0$ we consider a rational function $r_j \in R(K_j)$ with

$$\|f - r_j\|_{K_j} < \epsilon_1 := \frac{\epsilon}{1 + C}.$$

Next we find open sets V_j, W_j ($K_j \subset V_j$, $\overline{V}_j \subset W_j$) such that also $\|f - r_j\|_{\overline{W}_j} < \epsilon_1$, and we construct the C^∞-functions H_j corresponding to $\overline{V}_j$, W_j, according to Lemma 1. Then the function

$$s_j := f - (f - r_j)H_j$$

is continuous in $\mathbb{C}$, we have

$$s_j = \begin{cases} r_j & \text{in } V_j \supset K_j, \\ f & \text{outside } W_j, \end{cases}$$

and further

$$\|f - s_j\|_{\mathbb{C}} \leqslant \|f - r_j\|_{\overline{W}_j} < \epsilon_1.$$

Using these s_j, which are close to f, we construct analogously to (3.5)

$$(3.6) \qquad g_j(z) := s_j(z)\phi_j(z) + \frac{1}{\pi}\iint_{\mathbb{R}^2} \frac{s_j(\zeta)}{\zeta - z}(\bar{\partial}\phi_j)(\zeta)dm_\zeta \qquad (z \in \mathbb{R}^2).$$

It follows that

$$\|g_j - f_j\|_{\mathbb{C}} \leqslant \|f - s_j\|_{\mathbb{C}} + \|f - s_j\|_{\mathbb{C}} \cdot \frac{1}{\pi}\max_{z \in \mathbb{C}} \iint_{\mathbb{R}^2} \frac{|(\bar{\partial}\phi_j)(\zeta)|}{|\zeta - z|} dm_\zeta$$

$$< (1 + C)\epsilon_1 = \epsilon.$$

Moreover, by Theorem 1 c) and d), the function g_j is analytic outside t_{ϕ_j}, that is, in $\overline{U}_j^c$. It is also analytic in V_j, since $s_j = r_j$ is analytic in V_j. But we have $K \subset V_j \cup \overline{U}_j^c$, for if $z \in K$ and $z \notin \overline{U}_j^c$, then $z \in K \cap \overline{U}_j = K_j \subset V_j$. Hence g_j is analytic on K, and Runge's theorem guarantees a rational function $R_j \in R(K)$ such that $\|g_j - R_j\|_K < \epsilon$, and therefore $\|f_j - R_j\|_K < 2\epsilon$. Now each function f_j belongs to $R(K)$, and the proof of Theorem 3 is complete.

C_2. Applications of Bishop's theorem

With the help of Theorem 3 it is fairly straightforward to derive *two sufficient criteria* for $R(K) = A(K)$, which can easily be checked geometrically.

Theorem 4 (Mergelyan 1952). *Suppose the compact set $K \subset \mathbb{C}$ is such that the diameter of the components of K^c is $> \delta > 0$. Then $R(K) = A(K)$.*

This result is due to Mergelyan [1952, p. 317]. Garnett [1968, p. 463] observed that by means of the localization theorem it can be reduced to Mergelyan's theorem on polynomial approximation.

Proof. For an arbitrary point $z \in K$ we choose as U_z the disk with center at z and radius $\delta/2$, we write $K_z = K \cap \bar{U}_z$ and consider

$$K_z^c = K^c \cup \bar{U}_z^c.$$

This set contains all points outside $\bar{U}_z$. Moreover, if $\zeta \in K_z^c$ and $\zeta \in \bar{U}_z$, then $\zeta \in K^c$; hence ζ is in one of the components of K^c with diameter $> \delta$. This component must contain points outside $\bar{U}_z$, that is, in $K^c \subset K_z^c$ one can connect ζ with a point outside $\bar{U}_z$. Hence K_z^c is connected, and by Theorem 1 in §2 we have $f|_{K_z} \in P(K_z) \subset R(K_z)$. Since this holds for all $z \in K$, it follows that $f \in R(K)$, by Bishop's localization theorem.

Theorem 4 has the following consequence.

Corollary. *We have $R(K) = A(K)$ if K^c has only finitely many components.*

This corollary is our starting point for the second sufficient criterion. If each $z \in K$ has a neighborhood U_z such that U_z meets only finitely many components of K^c, then $K_z = K \cap \bar{U}_z$ itself has a complement K_z^c with finitely many components. It follows that $R(K_z) = A(K_z)$ for all $z \in K$, and thus $R(K) = A(K)$ by Bishop's theorem. Question: What can be said if there exist $z \in K$ for which each neighborhood U_z meets infinitely many components of K^c?

Let $M := \{z \in K$: each neighborhood U_z meets infinitely many components of $K^c\} \neq \phi$; obviously M is a closed subset of K.

Theorem 5 (Garnett). *If M is countable, then $R(K) = A(K)$.*

For the proof we mainly need Bishop's theorem, but we also require the following result, which is interesting in its own right.

Lemma 2. *Suppose f is continuous in $\mathbb{C}$ and analytic in an open set G; let $z_0 \in \mathbb{C}$. Then there exist functions f_n $(n = 1, 2, \ldots)$ such that:*

a) *f_n is continuous in $\mathbb{C}$;*

b) *f_n is analytic in $G \cup \{z_0\}$;*

c) $f_n(z) \Rightarrow f(z)$ $(n \to \infty; z \in \mathbb{C})$.

Proof. We may assume $z_0 = 0$; we choose auxiliary functions $g_n \in C^1(\mathbb{R}^2)$ such that

$$g_n(z) = \begin{cases} 1 & \text{for } |z| \leqslant 1/n, \\ 0 & \text{for } |z| \geqslant 2/n \end{cases}$$

and

$$|(\bar{\partial} g_n)(z)| \leqslant 2n \quad (z \in \mathbb{R}^2),$$

and we form the integral transforms (3.3):

$$G_n(z) := \frac{1}{\pi} \iint\limits_{\mathbb{R}^2} \frac{f(\zeta) - f(z)}{\zeta - z} (\bar{\partial} g_n)(\zeta) dm_\zeta \qquad (z \in \mathbb{R}^2).$$

All G_n are continuous in $\mathbb{C}$ and analytic in G, and by Theorem 1, e) we have

$$\|G_n\|_{\mathbb{R}^2} \leqslant 2 \cdot \frac{4}{n} \cdot 2 \max \{|f(\zeta) - f(0)| : |\zeta| \leqslant 2/n\} \cdot 2n \to 0 \qquad (n \to \infty).$$

Now the functions $f_n := f - G_n$ satisfy the conditions of the lemma, because, by (3.4),

$$f_n(z) = -\frac{1}{\pi} \iint\limits_{\mathbb{R}^2} \frac{f(\zeta)}{\zeta - z} (\bar{\partial} g_n)(\zeta) dm_\zeta \qquad (|z| < 1/n),$$

where the integration need only be carried out for $\{\zeta : 1/n < |\zeta| < 2/n\}$. Hence the functions f_n are analytic in $U_n = \{z : |z| < 1/n\}$.

Proof of Theorem 5. Let

$$N := \{z \in K: \text{there exists a neighborhood } U_z \text{ such that } f|_{K_z} \in R(K_z) \text{ for each } f \in A(K)\}.$$

Then N contains the "normal" points of K; note that the choice of U_z is to be independent of f. If $N = K$, then the assertion follows from Bishop's theorem. We set $S = K \setminus N$ and show that $S = \phi$.

i) S is closed in $\mathbb{C}$: Obviously N is open in K, hence S is closed in K and therefore also in $\mathbb{C}$;

ii) S is at most countable: If $z \notin M$, then $z \in N$; hence $S \subset M$, and the latter was assumed to be countable;

iii) S has no isolated points.

If we can show that iii) holds, we would be done, because a perfect set is uncountable or empty.

Suppose z_0 were an isolated point of S. Then there would exist a neighborhood U such that $\bar{U} \cap S = \{z_0\}$; all other points of K that lie in $\bar{U}$ would then belong to N. *We will show:* If $f \in A(K)$, then $f|_{K \cap \bar{U}} \in R(K \cap \bar{U})$. This would imply $z_0 \in N$, a contradiction to $z_0 \in S$.

To carry out our plan, we assume f has been extended continuously from K to $\mathbb{C}$, apply Lemma 2, and determine f_n such that $\|f - f_n\|_{\mathbb{R}^2} < \epsilon$. This function f_n belongs to $A(K)$ and is, in addition, analytic at z_0. Hence the localization theorem can be applied to $f_n|_{K \cap \bar{U}}$: The function f_n is analytic at z_0, and the points $z \in K \cap \bar{U}$ $(z \neq z_0)$ belong to N. Hence there exists a rational function $R \in R(K \cap \bar{U})$ such that $\|f_n - R\|_{K \cap \bar{U}} < \epsilon$. This implies $\|f - R\|_{K \cap \bar{U}} < 2\epsilon$, so that $f|_{K \cap \bar{U}} \in R(K \cap \bar{U})$, and this was to be shown.

D. Vitushkin's theorem; a report

This theorem gives a necessary and sufficient condition for $R(K) = A(K)$, where K is a compact set; hence it is analogous in the realm of rational approximation to Mergelyan's theorem. The preparations for the proof and the proof itself require considerable effort, which we must forgo in this introduction. The reader can find detailed presentations in Zalcman [1968] and Gamelin [1969, Chapter 8].

To formulate the result we need the *notion of the AC-capacity* of a set $M \subset \mathbb{C}$. Let $\mathfrak{R}(M)$ denote the class of functions f that are continuous in $\mathbb{C}$, satisfy $f(z) \to 0$ $(z \to \infty)$ and $\|f\|_{\mathbb{C}} \leqslant 1$, and are analytic in the complement of a compact subset of M. Consequently, at ∞ each $f \in \mathfrak{R}(M)$ has an expansion of the form

$$f(z) = c_1(f)/z + \text{decreasing powers.}$$

Then

$$\alpha(M) := \sup\{|c_1(f)| : f \in \mathfrak{R}(M)\}$$

is called the *AC-capacity of M*. It was introduced by Dolzhenko in 1962 to study rational approximation. Connections with other types of capacity are treated in the literature cited above.

Now we can state Vitushkin's theorem:

Theorem 6 (Vitushkin 1966). *We have $R(K) = A(K)$ if and only if*

$$(3.7) \qquad \alpha(K^c \cap D) = \alpha(K^{\circ c} \cap D), \quad \textit{that is,} \quad \alpha(D \setminus K) = \alpha(D \setminus K^\circ)$$

for each open disk D.

Other criteria, which also use the AC-capacity, are possible, too. Roughly, (3.7) means that the complements of K and of K° must be equally "thick" in the vicinity of each point, as measured by the AC-capacity α.

In the special case where K° is empty, (3.7) becomes $\alpha(D \setminus K) = \alpha(D)$; but the somewhat simpler analytic capacity γ can be used here instead of α (Vitushkin [1959]).

Remarks about §3

1. From the last-cited result by Vitushkin one can derive without difficulty the theorem of Hartogs and Rosenthal [1931] : If the compact set $K \subset \mathbb{C}$ has Lebesgue measure $\mu(K) = 0$, then $R(K) = C(K)$.

2. It is also worth mentioning that $R(K) = A(K)$ always implies $R(\partial K) = C(\partial K)$; this also is a criterion for rational approximation on sets without interior points. See Zalcman [1968, p. 74].

3. Note further that our Lemma 2 is only a special case of a more general result. For it is possible to choose the functions f_n in such a way that instead of satisfying b), the f_n are analytic on $G \cup \Gamma$, where Γ is a prescribed, twice continuously differentiable arc. The result is due to Vitushkin; see Gamelin [1969, Chapter 8.13].

4. Instead of $A(K)$ one can consider the subalgebra $A^\alpha(K) = \{f \in A(K) : f \in \text{Lip}\ \alpha \text{ on } K\}$ $(0 < \alpha \leqslant 1)$ and ask for conditions on K, under which each function $f \in A^\alpha(K)$ can be approximated by rational functions. Such conditions are known and use the notion of the analytic α-capacity of a set. See the survey article by Melnikov and Sinanjan [1976, §1 and §14].

§4. Roth's fusion lemma

This section also deals with approximation on compact sets. However, the fusion lemma by Roth [1976] serves mainly as a stepping stone to the study of approximation on noncompact sets, which will be taken up in the next chapter. In Part B we use the fusion lemma to prove Bishop's localization theorem anew.

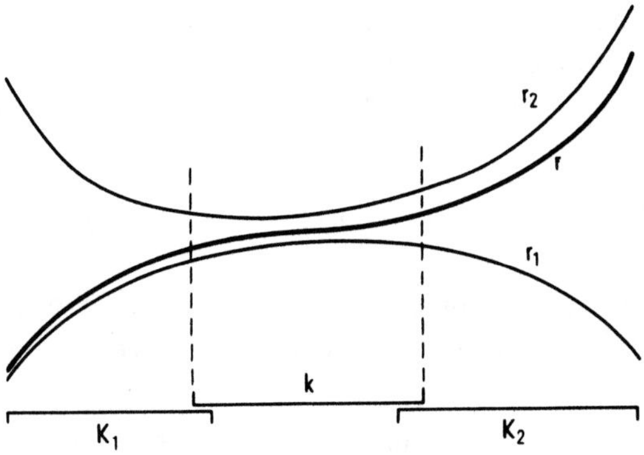

A. The fusion lemma

Suppose two rational functions r_1, r_2 are close together on a compact set k. The problem is to "connect" r_1 and r_2 in their further course by *a single* approximating rational function r; see the sketch above.

Theorem 1 (Roth 1976). *Suppose K_1, K_2, k are compact sets in the extended complex plane $\mathbb{C}$ such that $K_1 \cap K_2 = \phi$. Then there exists a constant A, depending only on K_1 and K_2, with the following property. If r_1, r_2 are rational functions such that*

$$|r_1(z) - r_2(z)| < \epsilon \qquad (z \in k),$$

then there exists a third rational function r such that

$$|r(z) - r_1(z)| < A\epsilon \qquad (z \in K_1 \cup k)$$

and

$$|r(z) - r_2(z)| < A\epsilon \qquad (z \in K_2 \cup k).$$

Note that nothing has been assumed about the location of the poles.

Note also that Theorem 1 follows immediately from Runge's theorem in the case where $K_1 \cap k = \phi$ or $K_2 \cap k = \phi$. Suppose, for example, that $K_1 \cap k = \phi$; we write

$$f(z) = \begin{cases} r_1(z) & (z \in K_1), \\ r_2(z) & (z \in K_2 \cup k), \end{cases}$$

let H_1 denote the sum of the principal parts of r_1 on K_1, and let H_2 denote the sum of the principal parts of r_2 on $K_2 \cup k$. Then $f - H_1 - H_2$ is analytic on $K_1 \cup K_2 \cup k$, and Runge's theorem furnishes a rational function R such that

$$|f(z) - H_1(z) - H_2(z) - R(z)| < \epsilon \quad (z \in K_1 \cup K_2 \cup k).$$

Thus the assertion of the theorem holds with $r = H_1 + H_2 + R$ and $A = 2$. Consequently, in Theorem 1 only the case where $K_1 \cap k \neq \phi$ and $K_2 \cap k \neq \phi$ is interesting.

Proof of Theorem 1. *1st step.* Preparations. a) It is enough to deal with the case $r_2 = 0$. For in the general case we let $\rho_1 = r_1 - r_2$, $\rho_2 = 0$, and there exists a rational function ρ such that

$$|\rho - \rho_1| < A\epsilon \ \text{ on } K_1 \cup k \quad \text{and} \quad |\rho| < A\epsilon \ \text{ on } K_2 \cup k;$$

that is,

$$|(\rho + r_2) - r_1| < A\epsilon \ \text{ on } K_1 \cup k$$

and

$$|(\rho + r_2) - r_2| < A\epsilon \ \text{ on } K_2 \cup k.$$

b) In addition, we can assume that $\infty \in K_2$, and we choose neighborhoods U_1, U_2 of K_1, K_2 such that $\overline{U}_1 \cap \overline{U}_2 = \phi$ and U_1, U_2 are bounded by finitely many Jordan curves. The diagram below shows a typical situation.

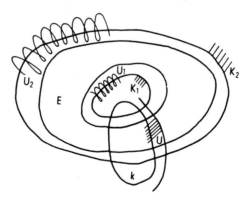

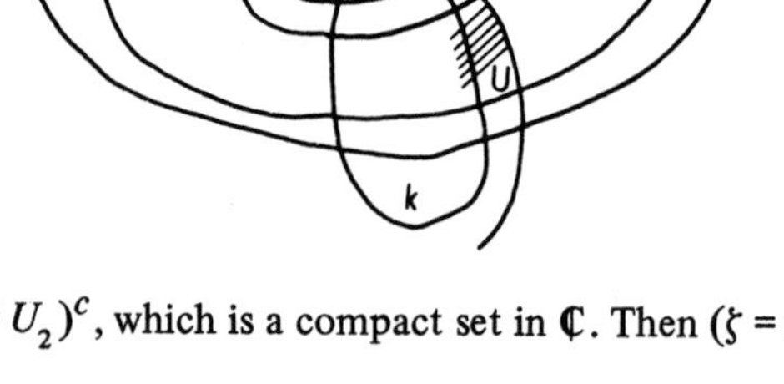

We set $E := (U_1 \cup U_2)^c$, which is a compact set in $\mathbb{C}$. Then ($\zeta = z + re^{i\phi}$)

$$\iint_E \frac{dm_\zeta}{|z - \zeta|} = \iint_E dr\, d\phi \leqslant 2\pi \operatorname{diam} E \quad (z \in \mathbb{C}).$$

c) Finally, we determine a function $H \in C^1(\mathbb{R}^2)$ with $0 \leqslant H(z) \leqslant 1$ $(z \in \mathbb{C})$ in such a way that

$$H(z) = 1 \quad \text{for } z \in \overline{U}_1 \quad \text{and} \quad H(z) = 0 \quad \text{for } z \in \overline{U}_2;$$

compare Lemma 1 in §3, B_2. For $z \in \mathbb{C}$ we obviously have

$$(4.1)\quad \frac{1}{\pi} \iint_E \frac{1}{|\zeta - z|} \, |(\overline{\partial} H)(\zeta)| \, dm_\zeta \leqslant \max_E |(\overline{\partial} H)(\zeta)| \cdot \frac{1}{\pi} \, 2\pi \operatorname{diam} E < A - 2,$$

where the constant A depends only on K_1, K_2.

2nd *step:* Construction of a function F meromorphic on $U_1 \cup U_2 \cup U$. For the proof, we refer back to the integral transform in §3, B_1, but with a function f that is *piecewise continuous* in the plane *except for poles.* It is defined as follows.

By assumption, there exists a neighborhood $U \supset k$ such that $|r_1(z)| < \epsilon$ for $z \in \overline{U}$. We set $f = r_1$ on $\overline{U} \cap E$ and extend this function continuously to all of E. Here $|f(z)| < \epsilon$ $(z \in E)$ is preserved (Tietze's theorem). On $E^c = U_1 \cup U_2$ we set $f = r_1$. If $P = \{\text{poles of } r_1 \text{ in } U_1 \cup U_2\}$, then f is piecewise continuous in $\mathbb{C} \setminus P$.

With this function f we define

$$(4.2)\quad \begin{aligned} F(z) &:= \frac{1}{\pi} \iint_{\mathbb{R}^2} \frac{f(\zeta) - f(z)}{\zeta - z} \, (\overline{\partial} H)(\zeta) dm_\zeta \\ &= \frac{1}{\pi} \iint_E \frac{f(\zeta) - f(z)}{\zeta - z} \, (\overline{\partial} H)(\zeta) dm_\zeta \qquad (z \in \mathbb{C} \setminus P). \end{aligned}$$

Considering the Pompeiu formula (§2, Theorem 2) for H, namely,

$$\begin{aligned} H(z) &= -\frac{1}{\pi} \iint_{\mathbb{R}^2} \frac{1}{\zeta - z} (\overline{\partial} H)(\zeta) dm_\zeta \\ &= -\frac{1}{\pi} \iint_E \frac{1}{\zeta - z} (\overline{\partial} H)(\zeta) dm_\zeta \qquad (z \in \mathbb{C}), \end{aligned}$$

one finds that

$$F(z) = f(z) H(z) + g(z) \qquad (z \in \mathbb{C} \setminus P),$$

where

$$g(z) = \frac{1}{\pi} \iint_E \frac{f(\zeta)}{\zeta - z} (\bar{\partial} H)(\zeta) dm_\zeta \qquad (z \in \mathbb{C}).$$

Concerning g we note: The function g is analytic in $E^c = U_1 \cup U_2$, and by (4.1) we have

$$|g(z)| < \epsilon(A - 2) \qquad (z \in \mathbb{C}).$$

Hence the function F defined in (4.2) has the following properties:

$H(z) = 1$ for $z \in U_1$; hence $F = f + g = r_1 + g$ is meromorphic in U_1 with the same poles as r_1;

$H(z) = 0$ for $z \in U_2$; hence $F = g$ is analytic in U_2;

F is analytic for $z \in U$, because in U we have that $f = r_1$ is rational and without poles; hence (4.2) is analytic in U.

Result: Except for finitely many poles in U_1, the function F is analytic in $U_1 \cup U_2 \cup U$.

3rd *step:* Estimates of F. For $z \in K_1$ we have $F - r_1 = (H - 1)r_1 + g = g$, so that $|F - r_1| < \epsilon(A - 2)$ $(z \in K_1)$; and for $z \in k$ we have $|F - r_1| \leqslant |r_1| + |g| < \epsilon + \epsilon(A - 2)$. Consequently

$$(4.3) \qquad |F(z) - r_1(z)| < \epsilon(A - 1) \qquad (z \in K_1 \cup k).$$

Analogously we have $F = g$ for $z \in K_2$, so that $|F(z)| < \epsilon(A - 2)$ $(z \in K_2)$, and for $z \in k$ we have $F = fH + g = r_1 H + g$, so that $|F(z)| < \epsilon + \epsilon(A - 2) = \epsilon(A - 1)$ $(z \in k)$. It follows that

$$(4.4) \qquad |F(z)| < \epsilon(A - 1) \qquad (z \in K_2 \cup k).$$

4th *step:* Construction of r. Since $K_1 \cup K_2 \cup k$ is compact in $U_1 \cup U_2 \cup U$ and F is analytic there except for finitely many poles, we can apply Runge's theorem to $F - \Sigma$ (where Σ contains the principal parts of F on U_1). As a result we obtain a rational function r such that

$$|F(z) - r(z)| < \epsilon \qquad (z \in K_1 \cup K_2 \cup k).$$

By (4.3) and (4.4), this function r satisfies

$$|r(z) - r_1(z)| < A\epsilon \qquad (z \in K_1 \cup k)$$

and

$$|r(z)| < A\epsilon \qquad (z \in K_2 \cup k),$$

as Theorem 1 asserts for $r_2 = 0$.

Remark. One can ask whether the following stronger form of the fusion lemma is true. Suppose K_1, K_2 are two compact sets in $\mathbb{C}$ with $k = K_1 \cap K_2 \neq \phi$. Assume that two rational functions r_1, r_2 are given with $|r_1(z) - r_2(z)| < \epsilon$ on k. Then there exists a rational function r such that

$$|r(z) - r_j(z)| < A\epsilon \quad \text{for } z \in K_j \qquad (j = 1, 2),$$

where A depends only on K_1 and K_2.

This version would imply Roth's lemma; however, it is *false in general.* This was shown originally by Gauthier using the "stitched disk" from §3, A_3. The set k on which K_1 and K_2 meet is a Jordan arc of positive Lebesgue measure. A simpler construction was given by Gaier [1983], where K_1 and K_2 are two squares with one common edge. Also studied in this paper are some cases in which the more general form of the fusion lemma *is* true.

B. A new proof of Bishop's theorem

As a first application of the fusion lemma we present, as indicated by Roth [1976, p. 108], an elementary proof of Bishop's localization theorem.

Suppose K is compact in $\mathbb{C}$, and suppose for each $z \in K$ there exists a disk U_z about z such that

$$f|_{K_z} \in R(K_z), \quad \text{where } K_z := \overline{U}_z \cap K.$$

We claim that $f \in R(K)$; i.e., f can be approximated uniformly on K by rational functions (with poles on K^c).

For $z \in K$, let u_z denote the disk about z whose radius is half that of U_z. Finitely many u_z cover K, say $u_{z_1}, u_{z_2}, \ldots, u_{z_N}$; let ρ denote their smallest radius.

We now cover the plane with a grid of mesh $h < \rho/3$; only finitely many squares meet K. Our choice of h implies: If a closed square Q with sides of length $2h$ (hence with diameter $\sqrt{8}\,h < 3h < \rho$) contains a point $P \in K$, then P lies in one of the u_{z_j}; hence $Q \subset U_{z_j}$ for this j. Consequently rational approximation is possible in each square Q with sides of length $2h$:

$$f|_{K \cap Q} \in R(K \cap Q).$$

To make approximation on larger rectangles possible, we use the fusion lemma. Suppose rational approximation is possible in a (closed) rectangle with sides of length, say, mh and $2h$ ($m \geqslant 2$).

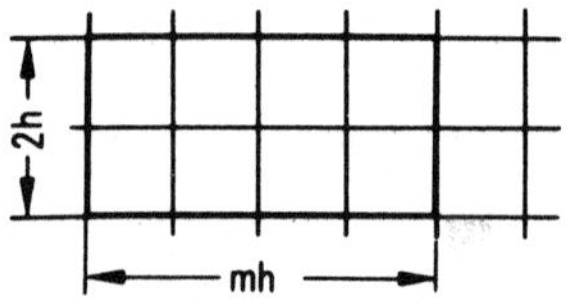

Let

R_1 denote the rectangle with sides of length $(m-1)h$, $2h$,
R the adjoining rectangle with sides of length h, $2h$, and
R_2 the rectangle with sides of length h, $2h$ adjoining R.

By assumption there exists for each $\epsilon > 0$ a rational function r_1 such that

$$|f(z) - r_1(z)| < \epsilon \quad \text{for } z \in (R_1 \cup R) \cap K.$$

In addition, rational approximation is possible in the square $R \cup R_2$:

$$|f(z) - r_2(z)| < \epsilon \quad \text{for } z \in (R \cup R_2) \cap K.$$

Now we apply the fusion lemma with

$$K_1 = R_1 \cap K, \qquad K_2 = R_2 \cap K, \qquad k = R \cap K;$$

note that

$$|r_1(z) - r_2(z)| < 2\epsilon \quad \text{for } z \in k.$$

The fusion lemma yields the existence of a rational function r such that

$$|r(z) - r_1(z)| < A \cdot 2\epsilon \quad \text{for } z \in (R_1 \cup R) \cap K$$

and

$$|r(z) - r_2(z)| < A \cdot 2\epsilon \quad \text{for } z \in (R_2 \cup R) \cap K.$$

This implies $|f(z) - r(z)| < (2A + 1)\epsilon$ for $z \in (R_1 \cup R \cup R_2) \cap K$; and since $\epsilon > 0$ was arbitrary, rational approximation is possible in the rectangle with sides of length $(m + 1)h$, $2h$. After finitely many steps we thus obtain

$$f|_{K \cap R^*} \in R(K \cap R^*)$$

for each rectangle R^* of height $2h$.

Finally, in a similar way, we apply the fusion lemma vertically and obtain that $f \in R(K)$.

Remark about §4

Although we have not considered the approximation of harmonic functions in this book, we point out that there exists a result for harmonic functions analogous to Roth's lemma; see Gauthier, Goldstein, and Ow [1980] for this and its consequences for the approximation of harmonic functions on closed sets.

Chapter IV

APPROXIMATION ON CLOSED SETS

So far, we have approximated functions defined on a compact set $K \subset \mathbb{C}$, and the approximating functions were polynomials or rational functions. Now we shall approximate functions defined on a set F that is closed in a domain G. Functions analytic or meromorphic in G will serve as approximating functions. In the special case where $G = \mathbb{C}$, one obtains approximation by entire functions. Here the rate of approximation (as $z \to \infty$) also plays a role. Several of these theorems can be used to construct analytic functions with complicated boundary behavior; we deal with these questions at the end of the chapter, in §5.

§1. Uniform approximation by meromorphic functions

Our first goal is the uniform approximation of functions on closed sets F by functions that are analytic in a domain $G \supset F$ (§2). The treatment becomes especially clear through the insertion of approximation by meromorphic functions. Roth [1938], [1973], [1976] and Nersesjan [1972] both deal with this topic.

A. Statement of the problem

Suppose $G \subset \mathbb{C}$ is an arbitrary domain and F is a relatively closed subset of G. Let $M(G)$ denote the set of functions meromorphic in G. They will be used for the uniform approximation of functions f on F.

It is easy to see that one can approximate more functions with functions in $M(G)$ than with rational functions. If, say, $G = \{z\colon |z| < 1\}$ and $F = \bigcup_2^\infty k_n$, where $k_n = \{z\colon |z - a_n| = r_n\}$, $a_n = 1 - 1/n$, $r_n = c/n^2$ ($c > 0$ such that the k_n are disjoint), then

$$f(z) = \Sigma_{n=2}^\infty \frac{c}{n^2} \cdot \frac{1}{(z - a_n)} \in M(G).$$

If there were a rational function r such that $|f(z) - r(z)| < 1$ ($z \in F$), then it would follow in particular that

$$|(z - a_n)f(z) - (z - a_n)r(z)| < r_n \qquad (z \in k_n).$$

If r has no pole inside k_n, then the inequality holds also inside k_n; and for $z = a_n$ we arrive at the contradiction

$$c/n^2 < r_n = c/n^2.$$

Hence r has a pole inside each circle k_n, and r is not rational.

B. Roth's approximation theorem

The following theorem reduces the problem of approximating f on F by meromorphic functions to the problem of approximating functions on compact sets by rational functions.

Theorem 1 (Roth 1976). *A function f can be uniformly approximated on F by functions in $M(G)$ without poles in F if and only if*

(1.1) $$f|_K \in R(K) \textit{ for each compact subset } K \subset F.$$

Remark. The following proof will show that the full condition (1.1) is not needed in order that f can be approximated on F by meromorphic functions. Rather, it is sufficient to assume that

(1.2) $$f|_K \in R(K) \quad \text{for } K = F \cap \overline{G}_n \qquad (n = 1, 2, \ldots),$$

where $\{G_n\}$ is some exhausting sequence of G with bounded domains G_n: $\overline{G}_n \subset G_{n+1}$, $\cup G_n = G$. Here $\overline{G}_n$ denotes the closure of G_n in $\mathbb{C}$.

Proof of Theorem 1. First, it is clear that (1.1) is *necessary* if f can be approximated on F by functions $m \in M(G)$ without poles in F. For each such m is analytic on K and, by Runge's theorem, can be approximated on K by rational functions. Hence $f|_K \in R(K)$.

Now suppose (1.1) is satisfied. Suppose the G_n are bounded domains such that $\overline{G}_n \subset G_{n+1}$ and $\cup G_n = G$; then $F_n := F \cap \overline{G}_{n+1}$ are compact subsets of F. Suppose further $\epsilon > 0$ and a monotone null sequence $\{\epsilon_n\}$ such that $\Sigma\, \epsilon_n < \epsilon/2$ are given.

For each $n = 1, 2, \ldots$ we now apply the fusion lemma from Chapter III, §4, with

$$K_1 = \overline{G}_n, \quad K_2 = \hat{\mathbb{C}} \setminus G_{n+1}, \quad \text{and } k = F_n.$$

Let A_n denote the constant A in the fusion lemma; we can assume $1 \leqslant A_n \uparrow$. As the two rational functions we use q_n and q_{n+1}, where

(1.3) $$|f(z) - q_n(z)| < \epsilon_n/2A_n \quad (z \in F_n),$$

and by assumption this condition can be satisfied by rational functions q_n (without poles on F_n). [*Note that* (1.1) *is used only for the sets* F_n.] Hence

$$|q_n(z) - q_{n+1}(z)| < \epsilon_n/A_n \qquad (z \in F_n).$$

By the fusion lemma there exists a rational function r_n such that

(1.4) $$|r_n(z) - q_n(z)| < \epsilon_n \quad \text{for } z \in \overline{G}_n \cup F_n$$

and

(1.5) $$|r_n(z) - q_{n+1}(z)| < \epsilon_n \quad \text{for } z \in (\hat{\mathbb{C}} \setminus G_{n+1}) \cup F_n.$$

With these rational functions r_n we write

$$m(z) := q_1(z) + \Sigma_{k=1}^{\infty} [r_k(z) - q_k(z)].$$

Now (1.4) implies that for fixed n and $z \in G_n$, the function $r_k - q_k$ is analytic in G_n as soon as $k \geqslant n$, and since (1.4) guarantees the uniform convergence of $\Sigma_{k \geqslant n}(r_k - q_k)$ in G_n, we see that m is analytic in G_n with the exception of finitely many poles. Hence m is meromorphic in G.

Finally, we show that f is approximated by m on F. First we have for $z \in F_1$

$$|m(z) - f(z)| \leqslant |q_1(z) - f(z)| + \Sigma_1^{\infty} |r_k(z) - q_k(z)| < \frac{\epsilon_1}{2A_1} + \Sigma_1^{\infty} \epsilon_k < \epsilon,$$

by (1.3) and (1.4). For $z \in F_n \setminus F_{n-1}$ (which is in $\hat{\mathbb{C}} \setminus G_k$ for $k = 1, 2, \ldots, n$), we have

$$|r_k(z) - q_{k+1}(z)| < \epsilon_k \qquad (k = 1, 2, \ldots, n-1)$$

by (1.5), and for $z \in F_k \supset F_n \setminus F_{n-1}$, we have

$$|r_k(z) - q_k(z)| < \epsilon_k \qquad (k \geqslant n)$$

by (1.4). If we now write

$$m - f = \Sigma_{k=1}^{n-1} (r_k - q_{k+1}) + (q_n - f) + \Sigma_{k=n}^{\infty} (r_k - q_k),$$

it follows immediately that

$$|m(z) - f(z)| \leqslant \Sigma_{k=1}^{n-1} \epsilon_k + \frac{\epsilon_n}{2A_n} + \Sigma_{k=n}^{\infty} \epsilon_k < \epsilon$$

for $z \in F_n \setminus F_{n-1}$. All together we have $|m(z) - f(z)| < \epsilon$ $(z \in F)$, and since (1.1) implies that f necessarily belongs to $A(F)$, the function m also cannot have poles on F.

C. Special cases of the approximation theorem

We now deal with three sufficient criteria for (1.1) or (1.2) to be satisfied. The third criterion uses some topological considerations, which we shall take up first.

C_1. The one-point compactification G^* of G; connectedness of $G^* \setminus F$

The *one-point compactification* G^* of a domain $G \subset \mathbb{C}$ is the extension of G, by the addition of an ideal point "∞", to a topological space $G^* = G \cup \{\infty\}$. A set $E \subset G^*$ is called open if

E is an open subset of G or if

$E = G^* \setminus K$ for some compact subset K of G.

With this topology, G^* is a compact space; see, for example, Taylor [1958, p. 67].

We now deal with the condition, which will come up again in §2, that the space $S = G^* \setminus F$ is connected.

Lemma. *The space $S = G^* \setminus F$ is connected if and only if each component Z of the open set $G \setminus F$ has an accumulation point on ∂G or is unbounded.*

Example. Suppose

$G = \{z\colon \operatorname{Re} z > 0\}$ and

$F = \{z\colon |z| = 1, \operatorname{Re} z > 0\}$

$\cup \{z = x\colon x > 1\}$

$\cup \bigcup_{n=3}^{\infty} \{z = re^{i\phi} : r > 1, \phi = \pi/n\}$.

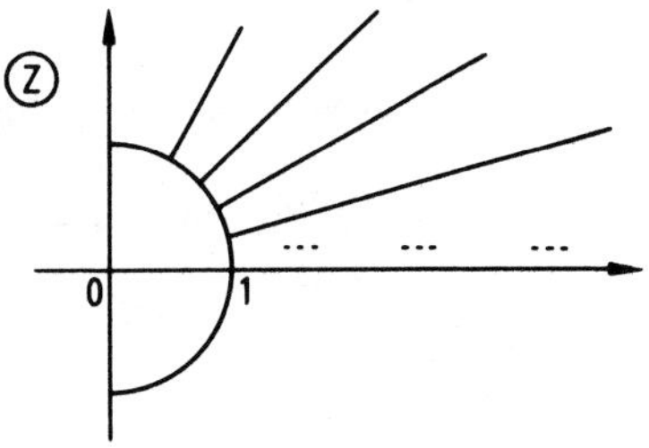

Here $S = G^* \setminus F$ is connected.

Proof. Suppose S is connected, and suppose there exists a bounded component Z of $G \setminus F$ such that $Z' \cap \partial G = \phi$. Then $\bar{Z} = Z \cup Z'$ is a compact subset of G; consequently $T := G^* \setminus \bar{Z}$ is open in G^*, so that $S \cap T$ is open in

S. In addition, Z is open in G, hence open in G^*, and therefore $Z \cap S = Z$ is open in S. Finally, we write

$$G \setminus F = Z \cup (G \setminus F) \cap (G \setminus \overline{Z}),$$

so that after addition of the ideal point ∞ we have

$$S = Z \cup (S \cap T);$$

this represents S as the union of two disjoint, nonempty, open subsets of S – contradiction.

For the converse, suppose $S = X \cup Y$, where X and Y are disjoint, nonempty, open subsets of S; suppose $\infty \in X$ and $y \in Y$. Then Y is bounded: We can write $X = S \cap T$, where T is open in G^* with $\infty \in T$, that is, $T = G^* \setminus K$ for some compact subset K of G. Consequently,

$$Y = S \setminus X = S \setminus (S \cap T) = S \setminus T \subset G^* \setminus T = K,$$

that is, $Y \subset K$. We now consider the component Z of $G \setminus F$ that contains y. Our assumption assures us that $Z \cap T \neq \phi$, so that also $Z \cap X \neq \phi$, and now

$$Z = (Z \cap X) \cup (Z \cap Y)$$

represents Z as the union of two disjoint, nonempty, open subsets of S. But then Z would not be connected.

Remark. In the general case, S decomposes into a component Z_∞, which contains ∞, and additional components Z that are compact in G. Here $Z_\infty = (\cup g) \cup \{\infty\}$, where the union is taken over the components g of $G \setminus F$ that have an accumulation point on ∂G or are unbounded.

C_2. Three sufficient criteria for meromorphic approximation

The situation is particularly simple if much is required of f.

Case 1: *f is analytic on F.*

By Runge's theorem, (1.1) is satisfied; hence f can be approximated on F uniformly by functions in $M(G)$. This case represents the extension of Runge's theorem to *closed*, not necessarily compact sets, and for $G = \mathbb{C}$ it dates back to Roth [1938, p. 105].

The next case represents an additional manageable and easily checked criterion for meromorphic approximation.

Case 2: *$f \in A(F)$ (f continuous on F and analytic in F°), and for each $z \in F$ there exists a disk U_z about z such that $(\bar{U}_z \cap F)^c$ is connected.*

According to the remark following Theorem 1, it is enough to verify the assumptions of Bishop's theorem for the compact sets

$$F_n := F \cap \bar{G}_n.$$

Here we take an exhausting sequence $\{G_n\}$ of G, where ∂G_n consists of finitely many Jordan curves. Thus for each $z \notin \bar{G}_n$ there exists a Jordan arc γ_z in the complement of $\bar{G}_n$, starting at z, such that diam $\gamma_z \geqslant \delta > 0$; $\delta = \delta_n$.

Now suppose $z \in F_n$, so that $z \in F$. We choose a disk V_z about z such that $V_z \subset U_z$ and diam $V_z < \delta$, and we assert that the open set $(\bar{V}_z \cap F_n)^c$ is connected. Here it is enough to consider $z_1 \in V_z$ and $z_1 \notin F_n$. Hence $z_1 \notin F$ or $z_1 \notin \bar{G}_n$. In the first case z_1 can, by assumption, be connected with ∂U_z without meeting F; hence z_1 can be connected with ∂V_z without meeting F_n. And if $z_1 \notin \bar{G}_n$, there exists a Jordan arc $\gamma_{z_1} \subset \bar{G}_n^c$ with diameter $\geqslant \delta$, which therefore meets ∂V_z. Hence z_1 can be connected with ∂V_z without meeting $\bar{G}_n$, that is, without meeting F_n. Hence $(\bar{V}_z \cap F_n)^c$ is connected, and by Mergelyan's theorem, approximation by polynomials is possible on $\bar{V}_z \cap F_n$.

The condition on F mentioned in Case 2 is satisfied in the following two examples.

Example 1: $G = \mathbb{C}$

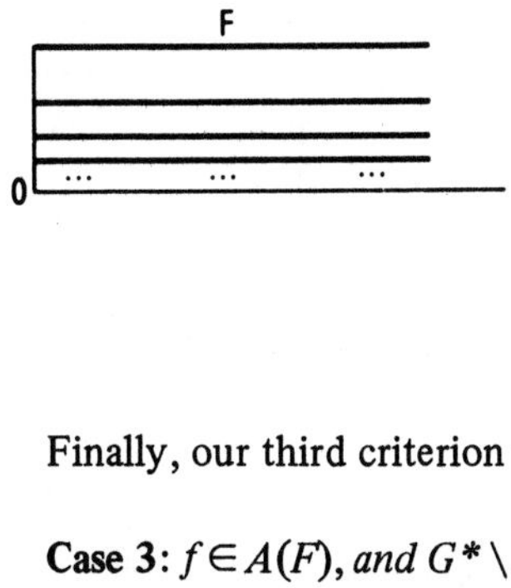

Example 2: $G = \{z: |z| < 1\}$

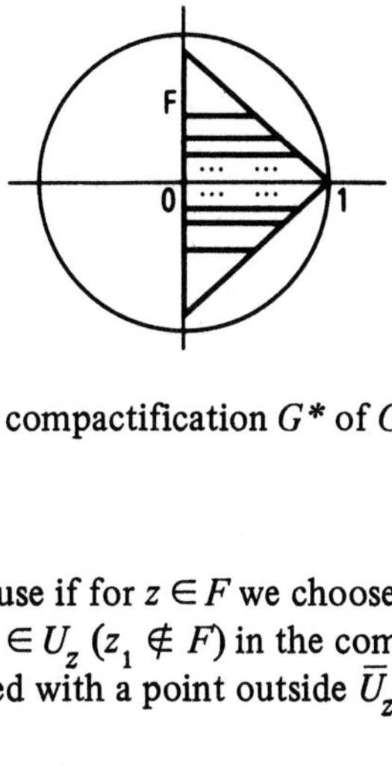

Finally, our third criterion uses the one-point compactification G^* of G.

Case 3: *$f \in A(F)$, and $G^* \setminus F$ is connected.*

Now the condition in Case 2 is satisfied. Because if for $z \in F$ we choose the disk U_z such that $\bar{U}_z \subset G$, then each point $z_1 \in U_z$ ($z_1 \notin F$) in the component of $G \setminus F$ that contains z_1 can be connected with a point outside $\bar{U}_z$ (Lemma). Hence $(\bar{U}_z \cap F)^c$ is connected in $\mathbb{C}$.

In Case 3 also, f can be approximated on F by meromorphic functions.

D. Characterization of the sets, where meromorphic approximation is possible

In Theorem 1, F was an arbitrary closed set in G, and the functions f were characterized that admit meromorphic approximation on F. Now we wish to characterize the sets F on which *every* function $f \in A(F)$ admits meromorphic approximation; as always, $A(F) = \{f: f$ continuous on F and analytic in $F^\circ\}$.

Theorem 2. *A relatively closed subset F of G has the property that every function $f \in A(F)$ can be uniformly approximated by functions in $M(G)$ if and only if*

$$R(F \cap \bar{g}) = A(F \cap \bar{g}) \tag{1.6}$$

for each domain g such that $\bar{g} \subset G$.

By Bishop's localization theorem it is enough to require (1.6) only for each disk k such that $\bar{k} \subset G$.

Proof. Condition (1.6) is sufficient. For if $f \in A(F)$ is given, so that $f \in A(F \cap \bar{g})$, then, by (1.6), f can be approximated by rational functions on $F \cap \bar{g}$, for each domain g such that $\bar{g} \subset G$. Hence (1.2) is satisfied.

To prove the necessity of (1.6), one must refer back to Vitushkin's theorem (Chapter III, §3). One shows for the AC-capacity α that

$$\alpha(k \setminus F) = \alpha(k \setminus F^\circ)$$

for each disk k. This can be established by using the definition of α and by using meromorphic approximation for a function $f \in A(F)$ that occurs there; the latter is possible by assumption. For details, see Nersesjan [1972, p. 406], and for the method of reasoning, see Zalcman [1968, p. 104].

Remark. In our theorems we have always required that the meromorphic functions m that approximate on F should not have poles on F. Analogous theorems are valid if the functions m are allowed to have poles on F; see Roth [1976, p. 110].

§2. Uniform approximation by analytic functions

Suppose again that $G \subset \mathbb{C}$ is an arbitrary domain and F is a relatively closed subset of G. Our problem is the following: Under what assumptions about F and G can each function $f \in A(F)$ be uniformly approximated on F by functions $g \in \mathrm{Hol}(G)$? Here

$$\mathrm{Hol}(G) = \{g: g \text{ analytic in } G\}.$$

The first theorem in this direction is due to Carleman (1927) and concerns the special case $G = \mathbb{C}$, $F = \mathbb{R}$; in other words, f is uniformly approximated on $\mathbb{R}$ by entire functions g. But Carleman even obtains "tangential approximation"; for this reason we postpone the theorem and a simple, direct proof to §3.

A. Moving the poles of meromorphic functions

With Runge's theorem we have already seen that it can be advantageous to relocate the poles of the approximating rational functions without affecting the approximation itself. Now it is important for us that an analogous result holds for meromorphic functions.

Theorem 1. *Suppose $G \subset \mathbb{C}$ is a domain, F is closed in G, and z_1, z_2 lie in the same component of $G \setminus F$. Then for each function m meromorphic in G and with pole at z_1 and for each $\epsilon > 0$, there exists a function m^* meromorphic in G that is analytic at z_1, has a pole at z_2, has no other poles except those of m, and for which*

$$|m(z) - m^*(z)| < \epsilon \qquad (z \in F). \tag{2.1}$$

Proof. We refer to the corresponding result for rational functions; see Theorem 3 in Chapter III, §1. The points z_1, z_2 can be connected by a Jordan arc γ in $G \setminus F$; hence $\gamma \cap F = \phi$. We write

$$m(z) = P(1/(z - z_1)) + H(z),$$

where P is a polynomial and H is analytic at z_1. According to the theorem just cited, there exists a polynomial Q such that

$$\left| P\left(\frac{1}{z - z_1}\right) - Q\left(\frac{1}{z - z_2}\right)\right| < \epsilon \qquad (z \in F).$$

If we now let

$$m^*(z) = Q(1/(z - z_2)) + H(z),$$

then m^* satisfies the conditions of the theorem.

The theorem implies that for the approximation of functions on F by functions in $M(G)$ we can in any case combine *finitely many* poles in a component of $G \setminus F$.

B. Preliminary topological remarks

A topological space S is called *locally connected at* $a \in S$ if for each neighborhood u of a there exists a connected set $Z \subset u$ that contains a as an interior point. See, for example, Newman [1951, p. 84 ff.].

We shall apply this definition to $S = G^* \setminus F$ and $a = \infty$, where G^* is the one-point compactification of G introduced in §1, C_1. We say that a continuous arc $\gamma \subset G$ starting at $z_0 \in G$ *connects* z_0 *with* ∞ *in* G if for any given compact set $K \subset G$ there is a point on γ after which γ does not meet K any more.

The following lemma characterizes the local connectedness of $G^* \setminus F$ at ∞ by properties that can be gathered from G. Here U, V are neighborhoods of ∞ in G^*.

Lemma. *The space $S = G^* \setminus F$ is locally connected at ∞ if and only if the following holds: For every neighborhood U of ∞ there exists a neighborhood $V \subset U$ of ∞ with the property that each point $z \in V \setminus F$, $z \neq \infty$, can be connected with ∞ in G by a continuous arc $\gamma \subset U \setminus F$.*

Roughly, this means: Each point of $G \setminus F$ located sufficiently "far away" can be connected with ∞ in $G \setminus F$ such that the arc does not have to come "far inward". If one considers the two examples in §1, C_2, one sees that $G^* \setminus F$ is locally connected at ∞ in Example 1, but not in Example 2.

Proof of the lemma. a) Suppose the condition of the lemma is satisfied, and u is a neighborhood of ∞ in S, that is, $u = U \cap S = U \setminus F$ for some neighborhood U of ∞ in G^*. We use the neighborhood $V \subset U$ of ∞ from the condition in the lemma and write

$$Z = \cup \{\gamma_z : z \in V \setminus F, z \neq \infty\} \cup \{\infty\};$$

hence Z is a connected set in S, and $Z \subset U \setminus F = u$. Further, $Z \supset V \setminus F = V \cap S$, and the latter is a neighborhood of ∞ in S.

b) Now suppose S is locally connected at ∞ and U is a neighborhood of ∞ in G^*. It is sufficient to show:

(*) There exists a neighborhood $V \subset U$ of ∞ such that each point $z_0 \neq \infty$ in $V \setminus F$ can be connected in $U \setminus F$ with a point that is arbitrarily close to ∞.

For then we construct $\{U_n\}$ such that $U_{n+1} \subset U_n$ and $\cap U_n = \{\infty\}$ (exhaustion of G!) and the corresponding V_n such that $V_{n+1} \subset V_n$, and in the obvious way we connect countably many arcs to constitute γ, which lies in $U \setminus F$ and connects z_0 in G with ∞.

In order to show (*), we let $u = U \cap S$ and choose a subset $Z \subset u$ that is connected in S and that contains the open set v with $\infty \in v$:

$$v \subset Z, \quad v = V \cap S = V \setminus F.$$

For this neighborhood $V \subset U$ of ∞ condition (*) holds. We choose $z_0 \in V \setminus F$, $z_0 \neq \infty$, and let g denote the component containing z_0 of the in G open set $(G \setminus F) \cap U$. We show that g has ∞ as an accumulation point; this will establish (*).

But we have

$$(G \setminus F) \cap U = g \cup R,$$

where R is open in G (possibly $R = \phi$) and where no point of R can be an accumulation point of g. We now include ∞:

$$U \setminus F = g \cup R', \qquad R' = R \cup \{\infty\} \neq \phi;$$

the left-hand side is $U \cap S = u$, and g is open in G, thus also open in G^*, and therefore open in S. The intersection with $Z \subset u$ yields

$$Z = (g \cap Z) \cup (R' \cap Z) = A \cup B,$$

where A is open in Z and $B \neq \phi$. Since Z is connected, A cannot be closed in Z. Hence B contains an accumulation point of A, and therefore g has an accumulation point on $R' = R \cup \{\infty\}$. This accumulation point cannot be in R (see above); hence ∞ is an accumulation point of g.

C. Arakeljan's approximation theorem

Our goal is now to approximate $f \in A(F)$ by functions $g \in \mathrm{Hol}(G)$ uniformly on F. Here the following two properties of F relative to G play a role:

(K_1) $G^* \setminus F$ is connected,

(K_2) $G^* \setminus F$ is locally connected at ∞.

(K_1) already appeared in §1, C_1 and was discussed there. By the way, $G^* \setminus F$ obviously is always locally connected at each point $a \neq \infty$.

The approximation of f will be achieved in the following way:

$f \in A(F)$	given function
$\downarrow$ (K_1)	
$m \in M(G)$	approximates f (§1, C_2, Case 3)
$\downarrow$ (K_2)	

$r + g$ approximates m (Theorem 2 below). Here $g \in \text{Hol}(G)$, r rational, poles $\notin F$

$\downarrow$ $(K_1) + (K_2)$

g approximates m.

C_1. Approximation of meromorphic functions by analytic functions

Here, obviously, a relocation of infinitely many poles is required, which is possible by assumption (K_2).

Theorem 2. *If $m \in M(G)$ has no poles on F and if F satisfies condition (K_2), then for each $\epsilon > 0$ there exist a rational function r with poles outside F and a $g \in \text{Hol}(G)$ such that*

$$|m(z) - (r(z) + g(z))| < \epsilon \qquad (z \in F).$$

If in addition (K_1) holds, one can choose $r = 0$.

The special case $G = \mathbb{C}$ was already treated by Roth [1938, p. 110].

Proof. First two preliminary remarks. a) The poles of m do not accumulate in G. By (K_2), all poles of m, with finitely many exceptions, can be connected with ∞ in $G \setminus F$. To begin with we assume:

(2.2) All poles of m can be connected with ∞ in $G \setminus F$;

specifically, suppose the pole z_k is connected with ∞ by γ_k.

By (K_2), the γ_k can be chosen in such a way that each compact subset of G meets only finitely many of them. To achieve this, we determine a sequence $\{U_n\}$ of neighborhoods of ∞ with $\cap U_n = \{\infty\}$ such that we can choose $U = U_n$ and $V = U_{n+1}$ in the preceding lemma ($n = 1, 2, \ldots$). We connect the finitely many $z_k \in U_{n+1} \setminus U_{n+2}$ with ∞ in $U_n \setminus F$ by γ_k; now all the γ_k are specified and have the cited property.

b) Suppose G_n are bounded domains such that $\overline{G}_n \subset G_{n+1}$ and $\cup G_n = G$. Each $\overline{G}_n$ meets only finitely many γ_k.

Finally, we choose $\epsilon_n > 0$ such that $\Sigma\, \epsilon_n < \epsilon$:

Step 1. Only finitely many γ_k meet $\overline{G}_1$. The poles of m on these γ_k are pushed outside $\overline{G}_1$ along γ_k. By Theorem 1 there exists an $m_1 \in M(G)$ such that

$$|m(z) - m_1(z)| < \epsilon_1 \qquad (z \in F).$$

Result: All poles of m_1 lie on arcs γ_k or on their terminal segments that lie outside $\overline{G}_1$ and therefore do not meet $\overline{G}_1 \cup F$.

Step 2. Only finitely many γ_k meet $\overline{G}_2$. The poles of m_1 on these γ_k are pushed outside $\overline{G}_2$ along γ_k. By Theorem 1 there exists an $m_2 \in M(G)$ such that

$$|m_1(z) - m_2(z)| < \epsilon_2 \qquad (z \in F \cup \overline{G}_1).$$

Result: All poles of m_2 lie on arcs γ_k or on their terminal segments that lie outside $\overline{G}_2$ and therefore do not meet $\overline{G}_2 \cup F$.

Step n. Analogously there exists an $m_n \in M(G)$ such that

$$\text{(2.3)} \qquad |m_{n-1}(z) - m_n(z)| < \epsilon_n \qquad (z \in F \cup \overline{G}_{n-1}),$$

and all poles of m_n can be connected with ∞ without meeting $\overline{G}_n \cup F$.

Now we consider

$$g(z) := \lim_{n\to\infty} m_n(z) = m_N(z) + \Sigma_{n=N}^{\infty} [m_{n+1}(z) - m_n(z)].$$

Because $n \geqslant N$, the m_n are analytic in G_N, and by (2.3) the series converges uniformly in $\overline{G}_N$. Hence $g \in \mathrm{Hol}(G_N)$ for each N, and therefore $g \in \mathrm{Hol}(G)$.

By (2.3), we have for $z \in F$ that

$$|g(z) - m(z)| = |m_1(z) - m(z) + \Sigma_{n=1}^{\infty} [m_{n+1}(z) - m_n(z)]| < \epsilon_1 + \Sigma_{n=2}^{\infty} \epsilon_n < \epsilon.$$

With this, Theorem 2 is proved with $r = 0$, provided (2.2) holds. This will be the case if F satisfies condition (K_1) in addition to (K_2); see the lemma in §1, C_1.

If only (K_2) is satisfied, we combine the principal parts of the finitely many exceptional poles (which cannot be connected with ∞ in $G \setminus F$) of m to form a function r, and we consider $m - r$, for which (2.2) is valid. This completes the proof of Theorem 2.

We now discuss two examples.

Example 1: $G = \{z: |z| < 1\}$

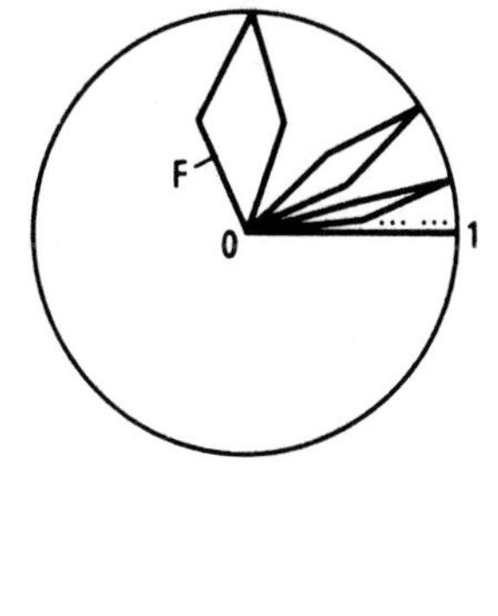

Example 2: $G = \{z: |z| < 1\}$

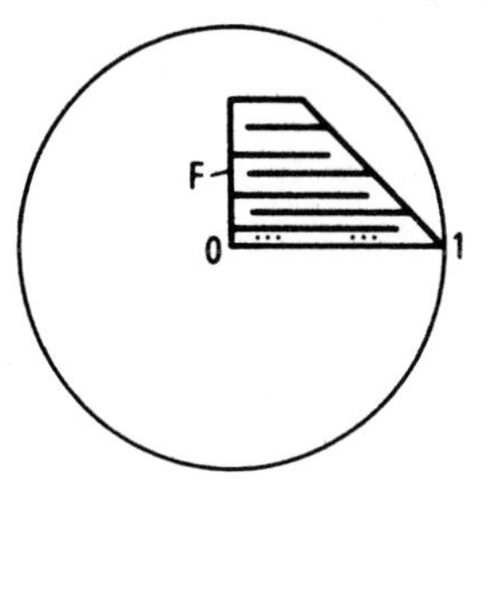

In Example 1, both conditions (K_1) and (K_2) are satisfied; in Example 2, (K_1) is satisfied, but (K_2) is not.

In the proof of Theorem 2, we used both conditions (K_1) and (K_2) for the step $r + g \Rightarrow g$. Whether (K_1) alone would be enough must remain open here. It would certainly be the case if for each point in $G \setminus F$ there exists an arc $\gamma \subset G \setminus F$, which from some point on lies outside each compact subset of G. Example 2 shows that this is not always the case, even if (K_1) is satisfied.

C_2. Arakeljan's theorem

We now give a complete answer to the question posed at the beginning of this section. We make use of the following definition.

Definition. *Suppose the set F is closed in G. Then F is called a Weierstrass set in G if each function $f \in A(F)$ can be approximated by functions in* Hol(G) *uniformly on F.*

With this definition we have the following result.

Theorem 3 (Arakeljan 1968). *The set F is a Weierstrass set in G if and only if the conditions*

(K_1) $\quad G^* \setminus F$ *is connected*

and

(K_2) $\quad G^* \setminus F$ *is locally connected at* ∞

are satisfied.

The case $G = \mathbb{C}$ was already dealt with by Roth in 1938, although with different terminology and only for sets F of two-dimensional measure 0. Works by Keldysh and Lavrentiev (1939) and Keldysh (1945) followed, in which F was a continuum such that $F^\circ = \phi$; and there was Mergelyan's report (1952). In these works (K_1) and (K_2) are often combined into one condition K_D (or K_G); K for Keldysh. Finally Arakeljan completely settled the case $G = \mathbb{C}$ in 1964 and the case of a general G in 1968. A careful exposition of the case $G = \mathbb{C}$ can be found in Fuchs [1968, pp. 9-34].

The approach used here – via meromorphic approximation – was indicated by Roth [1973], [1976]. It shows that in the final analysis, Theorem 3 can be derived from Mergelyan's theorem.

Proof of Theorem 3. In Theorem 2 in C_1 and in Case 3 in §1, C_2 we already proved that conditions (K_1) and (K_2) are sufficient.

We now show that (K_1) is necessary. If (K_1) were not satisfied, $G \setminus F$ would have a component Z that would be compact in G (lemma in §1, C_1); in particular, $\partial Z \subset G$, and therefore $\partial Z \subset F$. Let d denote the diameter of Z, choose $z_0 \in Z$, and consider $f(z) = 2d/(z - z_0)$, which is a function in $A(F)$.

By assumption there exists a $g \in \mathrm{Hol}(G)$ such that $|f(z) - g(z)| \leq 1$ $(z \in F)$, so that in particular $|f(z) - g(z)| \leq 1$ on ∂Z. Multiplication by $z - z_0$ yields

$$|2d - (z - z_0)g(z)| \leq d \qquad (z \in \partial Z).$$

By the maximum principle, this inequality also holds for $z \in Z$, and for $z = z_0$ we obtain a contradiction.

In order to show that (K_2) also is necessary, we observe first that (K_2) is equivalent to the following:

(*) For each neighborhood U of ∞ there exists a neighborhood $V \subset U$ of ∞ such that each point $z_0 \neq \infty$ of $V \setminus F$ can be connected in $U \setminus F$ with a point that is arbitrarily close to ∞.

Compare this with the reasoning in the proof of the lemma. If (*) does *not* hold, there exist a neighborhood $U = G \setminus K$ (K compact in G) of ∞ and a sequence of points $z_n \in G \setminus F$, $z_n \to \infty$, that cannot be connected in $U \setminus F$ with points arbitrarily close to ∞. Consequently the components g_n of $(G \setminus F) \cap U$ that contain z_n are compact in G.

The g_n can be assumed to be pairwise disjoint; let $d_n = \operatorname{diam} g_n$. Clearly, $\partial g_n \subset F \cup K$.

Using the Mittag-Leffler theorem, we now construct a function f meromorphic in G, which has simple poles at the points z_n with residues nd_n. Then $f \in A(F)$, and by assumption there exists a $g \in \mathrm{Hol}(G)$ such that $|f(z) - g(z)| \leq 1$ $(z \in F)$. In addition we have, of course, that $|f(z) - g(z)| \leq M$ $(z \in K)$ for some constant M.

Hence

$$|f(z) - g(z)| \leq \max(1, M) \quad \text{for } z \in \partial g_n,$$

and after multiplication by $z - z_n$ we have

$$|(z - z_n) f(z) - (z - z_n) g(z)| \leq d_n \cdot \max(1, M) \quad \text{for } z \in \partial g_n.$$

Since f has only the pole at z_n in g_n, the maximum principle can be applied again, and for $z = z_n$ we find that $nd_n \leq d_n \cdot \max(1, M)$, which is false for large n. This completes the proof of Theorem 3.

Remarks about §2

1. In the special case $G = \mathbb{C}$, the conditions (K_1) and (K_2) that occurred in Arakeljan's theorem can be combined as follows. We say a closed set $F \subset \mathbb{C}$ satisfies *condition* K if there exists a function $r(t)$ that is defined on $[0, \infty)$ such that $0 \leqslant r(t) \to \infty$ for $t \to \infty$ and that has the following property:

Each point $z \in F^c$ can be connected with ∞ in F^c by a Jordan arc γ_z that lies in $\{\zeta : |\zeta| \geqslant r(|z|)\}$.

This condition occurs first in Keldysh and Lavrentiev [1939, p. 746]. In the case $G = \mathbb{C}$, it is equivalent to $(K_1) + (K_2)$.

2. We also wish to point out newer developments that are based on Arakeljan's theorem. Above all, see the works by Stray [1974], [1977a], [1977b], [1978], [1980].

First, let $B_G(F) \subset A(F)$ denote the Banach algebra of functions *bounded* on F that admit uniform approximation on F by functions $g \in \mathrm{Hol}(G)$; further, let

$$A_u(F) := \{f \in A(F) : f \text{ is uniformly continuous on } F\} \subset A(F).$$

(i) If conditions (K_1) and (K_2) in Arakeljan's theorem are not both satisfied, then $A(F)$ decomposes into two subsets $A_G(F)$ and N: The functions in $A_G(F)$ can be uniformly approximated on F by functions in $\mathrm{Hol}(G)$, and the functions in N cannot. Stray shows: N even contains bounded functions, and he makes assertions about the functions in $A_G(F)$ and $B_G(F)$.

(ii) In addition, Stray proves a theorem for the class $A_u(F)$ that is analogous to Arakeljan's theorem. Each $f \in A_u(F)$ can be uniformly approximated by functions $g \in \mathrm{Hol}(G)$ whose restriction $g|_F$ belongs to $A_u(F)$ if and only if $G^* \setminus F$ is path-connected.

(iii) In [1977b], Stray generalizes the statement in (ii). Suppose E is a subset of $\partial G \cap \partial F$, and write

$$A_E(F) := \{f : f \text{ continuous on } E \cup F, \text{ analytic in } F^{\circ}\}.$$

If $E = \phi$, then $A_E(F) = A(F)$, and if $E = \partial G \cap \partial F$, one obtains $A_u(F)$ from above. Question: When can each function $f \in A_E(F)$ be uniformly approximated on F by functions $g \in \mathrm{Hol}(G)$ whose restriction $g|_F$ belongs to $A_E(F)$? This is the case if and only if $G^* \setminus F$ is path-connected and the set F_0 of "difficult to reach" boundary points of $G \setminus F$ is contained in E.

(iv) A generalization of Arakeljan's theorem to vector-valued analytic functions is discussed by Brown, Gauthier, and Seidel [1974].

§3. Approximation with given error functions

The theorems in the last two sections dealt with the *uniform* approximation of a function f on a closed set $F \subset G$ by functions $m \in M(G)$ or $g \in \text{Hol}(G)$. Since in general F is not compact in G, the question arises whether additional assertions can be made about the behavior of $|f(z) - m(z)|$ or $|f(z) - g(z)|$ for $z \in F$, $z \to \infty$, where ∞ is the ideal point of G^*.

The leading theorem here is due to Carleman (1927); we shall deal with it in an elementary way in Section A. Carleman already recognized the importance of his result for the study of the boundary behavior of analytic functions. Mainly for this reason the theory was further developed by Roth, Keldysh, Arakeljan, and Nersesjan.

A. The problem; Carleman's theorem

A_1. Tangential approximation; ϵ-approximation

First we define our problem. We assume G is an arbitrary domain and F is a relatively closed set in G.

Definition 1. *A function $\epsilon(z)$ that is defined on F and is positive and continuous there is called an error function. We say that $f \in A(F)$ admits ϵ-approximation on F by functions in $M(G)$ or* Hol(G) *if*

$$(3.1)\qquad \begin{aligned} &|f(z) - m(z)| < \epsilon(z) \qquad (z \in F) \text{ or} \\ &|f(z) - g(z)| < \epsilon(z) \qquad (z \in F) \end{aligned}$$

for some $m \in M(G)$ or $g \in$ Hol(G), *respectively.*

Definition 2. *We say that $f \in A(F)$ admits tangential approximation on F if for every error function $\epsilon(z)$ there exists an $m \in M(G)$ or a $g \in$* Hol(G), *for which* (3.1) *holds.*

Definition 3. *We say that F is a Carleman set in G if every function $f \in A(F)$ admits tangential approximation on F by functions in* Hol(G).

The corresponding notion for approximation by functions in $M(G)$ is not introduced. – Suppose a set F and a function $f \in A(F)$ are given; our task then consists of deciding whether ϵ-approximation or even tangential approximation is possible. First we turn to the special case $G = \mathbb{C}$, $F = \mathbb{R}$, which can be treated with elementary means, and later we admit general domains G and closed subsets F.

A_2. Two lemmas

We begin our preparation for Carleman's theorem with two lemmas.

Lemma 1. *Suppose* $G_k(a) := \mathbb{C} \setminus \{z = a + iy: |y| \geqslant 1/k\}$ $(a \in \mathbb{R}; k = 1, 2, \ldots)$. *Then there exist functions* $H_k^+(z, a)$ *and* $H_k^-(z, a)$ *that are analytic (and univalent) in* $G_k(a)$ *and have the properties:*

(a) $\quad |H_k^+(z, a)| < 1$ *and* $|H_k^-(z, a)| < 1$ *for* $z \in G_k(a)$;

and for each compact subset $K_1 \subset \{z: \text{Re } z < a\}$ *we have*

(b) $\quad H_k^+(z,a) \Rightarrow 0$ *and* $H_k^-(z, a) \Rightarrow 1$ *for* $z \in K_1$, $k \to \infty$,

whereas for each compact subset $K_2 \subset \{z: \text{Re } z > a\}$ *we have*

(c) $\quad H_k^+(z, a) \Rightarrow 1$ *and* $H_k^-(z, a) \Rightarrow 0$ *for* $z \in K_2$, $k \to \infty$.

We call H_k^+, H_k^- the "constriction functions" corresponding to the abscissa a. If one closes the shutter ∂G_k, then the H_k always provide a better approximation of the functions 0 and 1 on K_1 and K_2, respectively, and the H_k are *uniformly bounded* in G_k.

Proof. We assume $a = 0$ and start with the mapping $z = 2w/(1 - w^2)$. Its inverse $w = h(z)$ maps $\mathbb{C} \setminus \{z = iy: |y| \geqslant 1\}$ conformally onto $\{w: |w| < 1\}$. Here $0, \pm i$ are fixed points and $h'(0) > 0$. Obviously, $h(z) \to 1$ for $z \to \infty$ in $\text{Re } z > 0$, and $h(z) \to -1$ for $z \to \infty$ in $\text{Re } z < 0$.

Consequently, the functions $h_k(z) := h(kz)$ map the domains $G_k(0) = \mathbb{C} \setminus \{z = iy: |y| \geqslant 1/k\}$ conformally onto $\{w: |w| < 1\}$, and we have

$$h_k(z) \Rightarrow -1 \quad \text{for } k \to \infty, \text{ provided } z \in K_1,$$

and

$$h_k(z) \Rightarrow +1 \quad \text{for } k \to \infty, \text{ provided } z \in K_2.$$

Here K_1 and K_2 are compact subsets of $\{z: \text{Re } z < 0\}$ and $\{z: \text{Re } z > 0\}$, respectively. The functions

$$H_k^+ = \tfrac{1}{2}(1 + h_k) \quad \text{and} \quad H_k^- = \tfrac{1}{2}(1 - h_k)$$

thus have the properties mentioned in the lemma.

Lemma 1 is used in the proof of the following lemma.

Lemma 2. *Let* M_n *denote the set* $\{z: |z| \leqslant n\} \cup \{z = x: n \leqslant |x| \leqslant n + 1\}$ $(n \in \mathbb{N}$ *fixed), and suppose the function h is continuous on* M_n *and analytic in* $M_n^\circ = \{z: |z| < n\}$. *Then for each* $\epsilon > 0$ *there exists a polynomial P such that*

$$|h(z) - P(z)| < \epsilon \qquad (z \in M_n).$$

Of course, the assertion could be deduced immediately from Mergelyan's theorem, but we wish to give an elementary proof of this special case.

Proof. We write $M := M_n$ and show: There exists a function H analytic on M such that $|h(z) - H(z)| < \epsilon/2$ $(z \in M)$. Then Runge's theorem can be applied to H, which yields $|H(z) - P(z)| < \epsilon/2$ $(z \in M)$, and the assertion follows.

We write $M = K \cup I_1 \cup I_2$, where

$$K := \{z: |z| \leqslant n\}, \qquad I_1 := \{z = x: n \leqslant x \leqslant n+1\},$$

$$I_2 := \{z = x: -n-1 \leqslant x \leqslant -n\}.$$

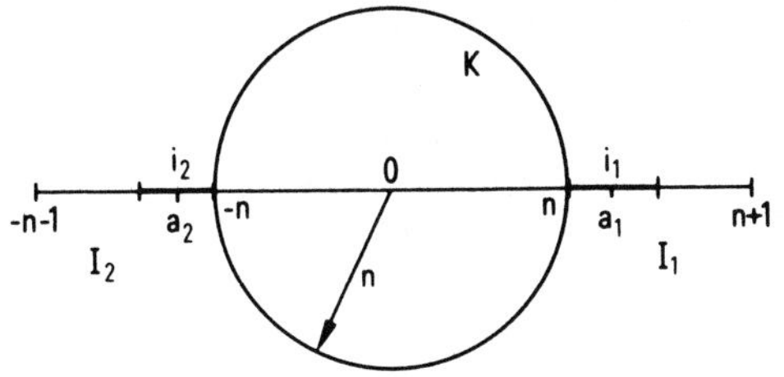

In addition, one can assume that $h(\pm n) = 0$; otherwise one would consider $h - L$ for some suitable linear function L. Now we choose $\epsilon' > 0$ and determine polynomials P_1, P_2, P_3 such that

$$|P_1 - h| < \epsilon' \quad \text{on } I_1, \qquad |P_2 - h| < \epsilon' \quad \text{on } I_2 \text{ (Weierstrass)}$$

and

$$|P_3 - h| < \epsilon' \quad \text{on } K \text{ (Fejér polynomials)}.$$

Then we determine the intervals

$$i_1 = \{z = x: n \leqslant x \leqslant n + \delta\} \text{ and } i_2 = \{z = x: -n - \delta \leqslant x \leqslant -n\}$$

such that

$$|h(x)| < \epsilon' \text{ and } |P_3(x)| < 2\epsilon' \quad \text{for } x \in i_1 \cup i_2.$$

Suppose a_1 and a_2 are the midpoints of i_1 and i_2, respectively. Next we fix $\delta > 0$ and set $c_1 := \max\{|P_1(x)|: x \in i_2\}$, $c_2 := \max\{|P_2(x)|: x \in i_1\}$.

Now the use of constriction functions, as we have constructed them in Lemma 1, is crucial. Specifically, let H_k^+ correspond to the abscissa a_1 and H_k^- to the abscissa a_2. With these two functions we construct

$$\Phi_k := P_1 H_k^+ + P_2 H_k^- + P_3(1 - H_k^+)(1 - H_k^-) \qquad (k = 1, 2, \ldots).$$

We observe:

(i) Each function Φ_k is analytic on the set M;

(ii) For $k \to \infty$ we have

$$H_k^+ \Rightarrow 1 \text{ and } H_k^- \Rightarrow 0 \text{ on } I_1 \setminus i_1; \text{hence } \Phi_k \Rightarrow P_1 \text{ on } I_1 \setminus i_1,$$

$$H_k^+ \Rightarrow 0 \text{ and } H_k^- \Rightarrow 1 \text{ on } I_2 \setminus i_2; \text{hence } \Phi_k \Rightarrow P_2 \text{ on } I_2 \setminus i_2,$$

$$H_k^+ \Rightarrow 0 \text{ and } H_k^- \Rightarrow 0 \text{ on } K; \text{hence } \Phi_k \Rightarrow P_3 \text{ on } K.$$

Due to the approximation properties of P_1, P_2, P_3 we have altogether that

$$|\Phi_k(z) - h(z)| < 2\epsilon' \quad \text{for } z \in M \setminus (i_1 \cup i_2),$$

as soon as k is sufficiently large.

But on i_1 we have

$$|\Phi_k - h| \leqslant |h| + |\Phi_k| < \epsilon' + |P_1| \cdot 1 + |P_2| \cdot |H_k^-| + |P_3| \cdot 4$$

for all k, where

$$|P_1| < 2\epsilon', |P_2| \leqslant c_2, |P_3| < 2\epsilon', \text{ and } |H_k^-| < \epsilon'/c_2,$$

as soon as k is sufficiently large; hence

$$|\Phi_k(z) - h(z)| < 12\epsilon' \quad \text{for } z \in i_1$$

for sufficiently large k. A corresponding inequality holds on i_2.

If originally one chooses $\epsilon' = \epsilon/24$, then for sufficiently large k the function Φ_k now yields a function H analytic on M, for which

$$|h(z) - H(z)| < \epsilon/2 \qquad (z \in M).$$

A_3. Carleman's theorem

Next we deal with tangential approximation in the special case $G = \mathbb{C}$, $F = \mathbb{R}$.

Theorem 1 (Carleman 1927). *For every function f continuous on* $\mathbb{R}$ *and every error function* $\epsilon(x)$ $(x \in \mathbb{R})$ *there exists an entire function g such that*

$$|f(x) - g(x)| < \epsilon(x) \qquad (x \in \mathbb{R}).$$

This theorem is the beginning of a series of theorems that will be discussed later. Carleman [1927] already generalized this theorem by replacing $\mathbb{R}$ by more general curves and systems of curves. See also the remarks at the end of §3.

For the proof it would seem reasonable to exhaust $\mathbb{R}$ by intervals $I_n = [-n, +n]$ and to start with polynomials P_n for which $\max\{|P_n(x) - f(x)|: x \in I_n\}$ is small. Then we would have $P_n(x) \Rightarrow f(x)$ $(n \to \infty)$ on each compact subset of $\mathbb{R}$. But $P_n(x) \Rightarrow f(x)$ $(n \to \infty)$ is possible on all of $\mathbb{R}$ only if f itself is a polynomial; the idea for the proof of Theorem 1 must be refined. We follow the proof given by Kaplan [1955, pp. 43–44], who attributes it to Brelot.

Proof of Theorem 1. First we choose a null sequence $\{\delta_n\}$ such that

$$0 < \delta_{n+1} < \delta_n, \quad \delta_n < \max\{\epsilon(x): n \leqslant |x| \leqslant n+1\} \quad (n = 0, 1, 2, \ldots),$$

and we write

$$\epsilon_n := \delta_{n+1} - \delta_{n+2} \quad (n = 0, 1, 2, \ldots) \text{ and } \epsilon_{-1} := 0.$$

In the 0th step we approximate f on $[-1, +1]$ by a polynomial P_0 within ϵ_0:

$$|P_0(x) - f(x)| < \epsilon_0 \quad \text{for } |x| \leqslant 1.$$

In the 1st step we start by writing

$$h_1(z) = \begin{cases} P_0(z) & \\ f(x) + (2 - x)[P_0(1) - f(1)] & \text{for} \\ f(x) + (2 + x)[P_0(-1) - f(-1)] & \end{cases} \begin{cases} \{z: |z| \leqslant 1\} \\ \{x: 1 \leqslant x \leqslant 2\} \\ \{x: -2 \leqslant x \leqslant -1\}. \end{cases}$$

Then h_1 is continuous on M_1, analytic in M_1°, and Lemma 2 yields a polynomial P_1 such that $|P_1(z) - h_1(z)| < \epsilon_1$ $(z \in M_1)$; hence in particular

$$|P_1(z) - P_0(z)| < \epsilon_1 \quad \text{for } |z| \leqslant 1$$

and $|P_1(x) - h_1(x)| < \epsilon_1$ for $1 \leqslant |x| \leqslant 2$. It follows that $|P_1(x) - f(x)| < \epsilon_1$ for $x = \pm 2$, and $|P_1(x) - f(x)| < \epsilon_1 + \epsilon_0$ for $1 \leqslant |x| \leqslant 2$.

Analogously, in the 2nd step we start with

$$h_2(z) = \begin{cases} P_1(z) & \\ f(x) + (3 - x)[P_1(2) - f(2)] & \text{for} \\ f(x) + (3 + x)[P_1(-2) - f(-2)] & \end{cases} \begin{cases} \{z\colon |z| \leqslant 2\} \\ \{x\colon 2 \leqslant x \leqslant 3\} \\ \{x\colon -3 \leqslant x \leqslant -2\}, \end{cases}$$

and Lemma 2 yields a polynomial P_2, for which

$$|P_2(z) - P_1(z)| < \epsilon_2 \quad \text{for } |z| \leqslant 2$$

as well as $|P_2(x) - f(x)| < \epsilon_2$ for $x = \pm 3$ and $|P_2(x) - f(x)| < \epsilon_2 + \epsilon_1$ for $2 \leqslant |x| \leqslant 3$.

In the nth step we obtain a polynomial P_n such that

$$|P_n(z) - P_{n-1}(z)| < \epsilon_n \quad \text{for } |z| \leqslant n$$

and

$$|P_n(x) - f(x)| < \epsilon_n \qquad \text{for } x = \pm(n+1) \text{ and}$$

$$|P_n(x) - f(x)| < \epsilon_n + \epsilon_{n-1} \quad \text{for } n \leqslant |x| \leqslant n+1.$$

The last two inequalities also hold for $n = 0$ (because $\epsilon_{-1} = 0$).

If we now write

$$g(z) := \lim_{n\to\infty} P_n(z) = P_0(z) + \Sigma_{k=0}^{\infty} [P_{k+1}(z) - P_k(z)],$$

it follows that g is an entire function, because the series of polynomials converges uniformly on compact subsets of $\mathbb{C}$.

If $x \in \mathbb{R}$ and therefore $n \leqslant |x| < n + 1$, we have in addition that

$$g(x) - f(x) = [g(x) - P_n(x)] + [P_n(x) - f(x)],$$

where $|P_n(x) - f(x)| < \epsilon_n + \epsilon_{n-1}$ and

$$|g(x) - P_n(x)| = |\Sigma_{k\geqslant n} [P_{k+1}(x) - P_k(x)]| < \Sigma_{k\geqslant n} \epsilon_{k+1}.$$

Altogether we obtain

$$|g(x) - f(x)| < \Sigma_{k=n-1}^{\infty} \epsilon_k = \begin{cases} \delta_n & \text{for } n > 0, \\ \delta_1 < \delta_0 & \text{for } n = 0; \end{cases}$$

hence $|g(x) - f(x)| < \delta_n < \epsilon(x)$ for each $x \in \mathbf{R}$.

B. The special case where F is nowhere dense

We now turn to general domains G and closed subsets F; here the case $F^{\text{o}} = \phi$ is particularly easy to dispatch. We rely on a method of proof that seems to be due to Arakeljan [1964b]; it is also useful in the case $F^{\text{o}} \neq \phi$.

B_1. Sufficient conditions for ϵ-approximation

Suppose F is closed in G and that *uniform* approximation on F, by functions in Hol(G) or $M(G)$, is always possible; initially we do not require that $F^{\text{o}} = \phi$. We show: For certain error functions ϵ-approximation is even possible.

Lemma 3. *Suppose F is a Weierstrass set in G and $\psi \in A(F)$. Then every function $f \in A(F)$ admits ϵ-approximation by functions in* Hol(G) *for* $\epsilon(z) = |e^{\psi(z)}|$.

Proof. Since F is a Weierstrass set, there exists a $g_1 \in \text{Hol}(G)$ such that $|\psi(z) - g_1(z)| < 1$ $(z \in F)$. We let $h = e^{g_1 - 1} \in \text{Hol}(G)$, consider $f/h \in A(F)$, and determine a $g_2 \in \text{Hol}(G)$ such that

$$|(f/h)(z) - g_2(z)| < 1 \qquad (z \in F).$$

It follows that

$$|f(z) - h(z)g_2(z)| < |h(z)| = \exp\{\text{Re}\, g_1(z) - 1\}$$
$$< \exp\{\text{Re}\, \psi(z)\} = |e^{\psi(z)}| \ (z \in F),$$

as was asserted.

Even easier is the proof of the following remark, which will be used later.

Remark. *Suppose F is a Weierstrass set in G and*

(3.2) *suppose there exists $H \in$* Hol(G) *such that* $0 < |H(z)| \leqslant \epsilon(z) \qquad (z \in F)$.

Then every function $f \in A(F)$ admits ϵ-approximation by functions in Hol(G).

To see this, one simply considers f/H and applies the Weierstrass property of F (once!).

The result is even smoother than Lemma 3 if one studies ϵ-approximation by meromorphic functions.

Lemma 4. *Suppose F is such that every function $f \in A(F)$ admits uniform approximation on F by functions in $M(G)$. Suppose $h \in A(F)$ with $0 < |h(z)| < 1$ $(z \in F)$. Then every function $f \in A(F)$ admits ϵ-approximation by functions in $M(G)$ for $\epsilon(z) = |h(z)|$.*

Proof (Nersesjan [1972, p. 411], Roth [1976, p. 109]). First we approximate $2/h \in A(F)$ uniformly by functions in $M(G)$:

$$\left|\frac{2}{h(z)} - m_1(z)\right| < 1 \qquad (z \in F)$$

for $m_1 \in M(G)$. This implies

$$|m_1(z)| > \frac{2}{|h(z)|} - 1 > \frac{1}{|h(z)|}, \quad \text{so that} \quad \frac{1}{|m_1(z)|} < |h(z)| \qquad (z \in F);$$

in particular, m_1 has no zero on F.

Further, $m_1 f \in A(F)$ can be uniformly approximated by functions in $M(G)$:

$$|m_1(z)f(z) - m_2(z)| < 1 \qquad (z \in F)$$

for $m_2 \in M(G)$. Letting $m = m_2/m_1 \in M(G)$, we obtain from this

$$|f(z) - m(z)| < 1/|m_1(z)| < |h(z)| \qquad (z \in F),$$

as was asserted.

In Section B_2 we shall apply these lemmas to sets with $F^{\circ} = \phi$. First we present two theorems where $G = \mathbb{C}$ but where $F^{\circ} = \phi$ is not required.

Theorem 2. *Suppose $F \subset \mathbb{C}$ is such that every function $f \in A(F)$ admits uniform approximation by functions in $M(\mathbb{C})$. Suppose further that $f \in A(F)$, $\epsilon > 0$, and $n \in \mathbb{N}$ are given. Then there exists a meromorphic function $m \in M(\mathbb{C})$ such that*

$$(3.3) \qquad |f(z) - m(z)| < \epsilon \qquad (z \in F)$$

and

$$(3.4) \qquad |f(z) - m(z)| = O(|z|^{-n}) \qquad (z \in F; z \to \infty).$$

In other words, in addition to uniform approximation one can achieve approximation at the rate given by (3.4) without additional restrictions on F.

Proof (Roth [1976, p. 109]). For $F = \mathbb{C}$ the theorem is trivial. Thus we can assume that $z_0 \in \mathbb{C} \setminus F$ and that $\eta > 0$ is chosen such that $|z - z_0|^n > \eta$ for $z \in F$. We apply Lemma 4 with

$$h(z) = \epsilon\eta(z - z_0)^{-n}$$

and immediately obtain (3.3) and (3.4).

Note that the meromorphic function m depends on n. In general, (3.4) does not hold for all $n \in \mathbb{N}$ and fixed $m \in M(\mathbb{C})$. For example, if $F = \{z: |z| \geq 1\}$, then (3.4) holds for fixed f and m and all $n \in \mathbb{N}$ only if f itself is meromorphic in $\mathbb{C}$; thus (3.4) does not hold for $f(z) = e^{1/z} \in A(F)$, say.

The following result concerning approximation on Weierstrass sets by entire functions can be obtained by elementary methods.

Theorem 2′. *Suppose F is a Weierstrass set in $\mathbb{C}$, $f \in A(F)$, and $\epsilon > 0$. Then there exists an entire function g such that*

$$|f(z) - g(z)| < \epsilon \quad \textit{and} \quad |f(z) - g(z)| < 1/|z| \qquad (z \in F).$$

With stronger auxiliary results it is possible to improve this theorem significantly (see §4, A), but the present version is sufficient for our later applications. For example, it allows us immediately to construct entire functions g that are bounded outside "small sets."

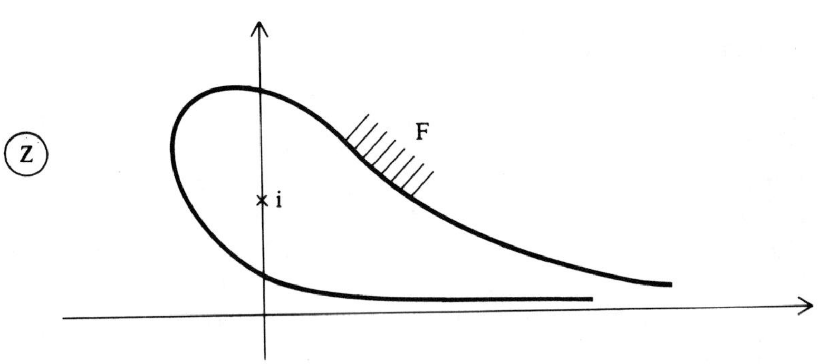

Suppose we take $f(z) = 1/(z - i)$, say. Then, in the example sketched above, one immediately obtains an entire function g such that

$$\lim_{r\to\infty} g(re^{i\phi}) = 0 \quad \text{for all } \phi \in [0, 2\pi].$$

Proof. For $F = \mathbb{C}$ the statement is trivial. Therefore, suppose $F \neq \mathbb{C}$ and k is a closed disk in F^c with center z_0. We write

$$F^* = F \cup k, \qquad f^* = \begin{cases} f & \text{on } F, \\ 0 & \text{on } k. \end{cases}$$

Then F^* is again a Weierstrass set and $f^* \in A(F^*)$. Thus, for arbitrary $\alpha > 0$, there exists an entire function g^* such that

$$\left| \frac{z - z_0}{\alpha} f^*(z) - g^*(z) \right| < 1 \qquad (z \in F^*).$$

In particular, we have $|g^*(z_0)| < 1$ and therefore

$$\left| \frac{z - z_0}{\alpha} f^*(z) - (g^*(z) - g^*(z_0)) \right| < 2 \qquad (z \in F^*);$$

consequently

$$\left| f^*(z) - a \frac{g^*(z) - g^*(z_0)}{z - z_0} \right| < \frac{2\alpha}{|z - z_0|} \qquad (z \in F^*).$$

On F we then have

$$|f(z) - g(z)| < \frac{2\alpha}{|z - z_0|} < \min(\epsilon, 1/|z|),$$

assuming α was chosen sufficiently small.

Remark. In general, there is no analog to Theorem 2′ for Weierstrass sets in other domains G. For example, if $G = \{z: |z| < 1\}$ and $F = G \cap \{z: \operatorname{Re} z \geqslant 0\}$, then F is a Weierstrass set in G, but the relation $|f(z) - g(z)| \to 0$ for $|z| \to 1$ $(z \in F)$ can hold for $f \in A(F)$, $g \in \operatorname{Hol}(G)$ only if $f = g$. – Conditions on F under which $|f(z) - g(z)| \to 0$ for $|z| \to 1$ $(z \in F)$ can be achieved are given by Brown, Gauthier, and Seidel [1975, p. 4]. For example, it is sufficient that $\overline{F} \cap \partial G$ has linear measure 0, but "noodles" are also permitted [1975, p. 5].

B_2. Tangential approximation if $F^{\circ} = \phi$

If $F^{\circ} = \phi$, Lemmas 3 and 4 allow us to reduce the problem of tangential approximation to that of uniform approximation; the latter was completely solved in §1 and §2.

Theorem 3. *Suppose F is closed in G and $F^{\circ} = \phi$.*

a) *F is a Carleman set in G if and only if F is a Weierstrass set in G.*

b) *Every function $f \in A(F)$ admits tangential approximation on F by functions in $M(G)$ if and only if every function $f \in A(F)$ admits uniform approximation on F by functions in $M(G)$.*

For $G = \mathbb{C}$ and $F^{\circ} = \phi$, the characterization of Carleman sets is due to Mergelyan [1952, pp. 327–329] who improved a method of proof first employed by Keldysh and Lavrentiev (1939; there F is a continuum) by using his main theorem. (Mergelyan requires the property "F a continuum" but does not need it.) For general G and $F^{\circ} = \phi$, the characterization is due to Arakeljan [1968]. Statement b) was proved by Nersesjan [1972] and Roth [1973].

The **proof** of Theorem 3 is obvious. If $\epsilon(z)$ is an arbitrary error function, we apply Lemma 3 with $\psi(z) = \log \epsilon(z)$ and Lemma 4 with $h(z) = \min(\epsilon(z), 1/2)$ (this can be done because $F^{\circ} = \phi$). Hence every $f \in A(F)$ admits tangential approximation if every $f \in A(F)$ admits uniform approximation.

Remark. The proof remains valid even for $F^{\circ} \neq \phi$ if we limit ourselves to ϵ-approximations whose error functions have the property

(3.5) $\qquad \epsilon(z)$ is constant on each component of F°.

This type of error function plays a role in the work by Brown and Gauthier [1973].

C. Nersesjan's theorem

We now deal with tangential approximation on sets F that do not satisfy $F^{\circ} = \phi$. No general theorem is known for approximation by functions in $M(G)$, in contrast to approximation by analytic functions.

C_1. Condition (A); a lemma

Tangential approximation on F will be possible only with an additional assumption.

Definition 4. *We say F satisfies condition (A) if for every compact subset $K \subset G$ there exists a neighborhood V of ∞ in G^* such that no component of F° intersects both K and V.*

This condition was first introduced for $G = \{z\colon |z| < R\}$ $(0 < R \leqslant \infty)$ by Gauthier [1969]. If (A) is satisfied, then all components of F° lie necessarily compact in G; but this condition is not sufficient (Example 2).

Example 1:
$G = \mathbb{C}$

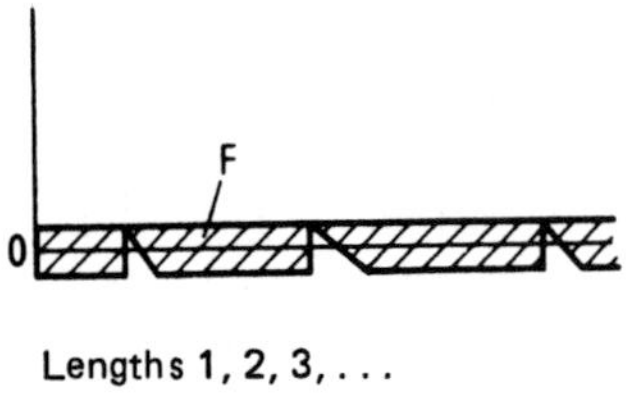

(A) satisfied

Example 2:
$G = \mathbb{C}$

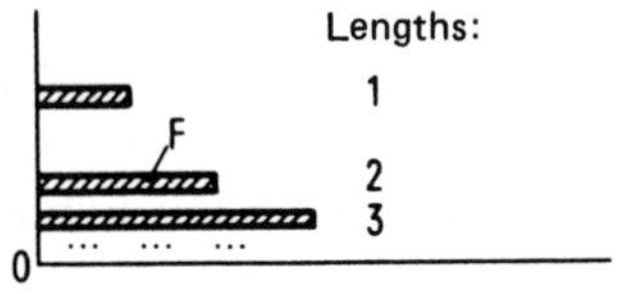

(A) violated

Example 3:
$G = \{z\colon |z| < 1\}$

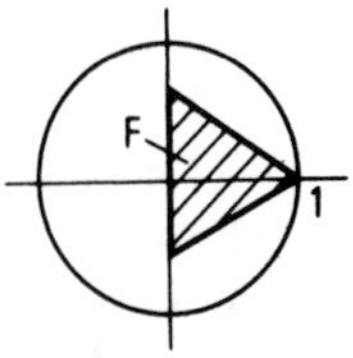

(A) violated

Roughly speaking, (A) requires that "long islands" of F must move out to ∞.

In Section C_2 we need the following lemma.

Lemma 5. *Suppose h is analytic in $\{z\colon |z| \leqslant r\}$ and $0 < r' < r$. Suppose further that there exists a sequence of Jordan arcs γ_n that connect $\{z\colon |z| = r\}$ with $\{z\colon |z| = r'\}$ and on which*

$$|h(z)| \leqslant \epsilon_n \qquad (z \in \gamma_n),$$

where $\{\epsilon_n\}$ is a null sequence. Then $h = 0$.

Proof. We can assume

$$M := \max\{|f(z)|\colon |z| = r\} \leqslant 1.$$

Let $\alpha_n(z)$ denote the harmonic measure of γ_n with respect to $\{z\colon |z| < r\} \setminus \gamma_n$; we consider $\alpha_n(z)$ on the disk $k = \{z\colon |z| \leqslant r'/2\}$. Then

$$\alpha_n(z) \geqslant \alpha > 0 \qquad (z \in k; n = 1, 2, \ldots),$$

and by the "two-constants theorem,"

$$|h(z)| \leqslant \epsilon_n^{\alpha_n(z)} \cdot M^{1-\alpha_n(z)} \leqslant \epsilon_n^{\alpha_n(z)} \leqslant \epsilon_n^{\alpha} \qquad (z \in k;\ n = 1, 2, \ldots).$$

Hence $h = 0$ in k, and therefore $h = 0$ in $\{z: |z| < r\}$.

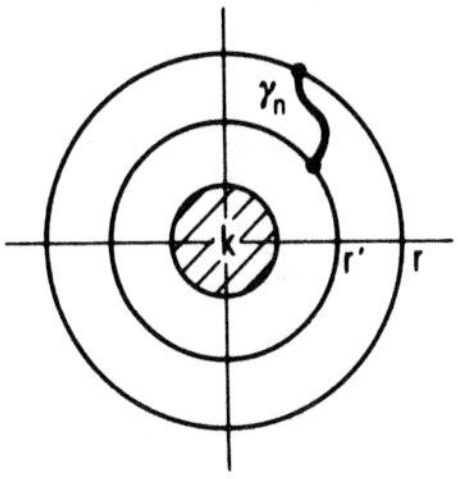

C_2. Nersesjan's theorem

Now we answer the question: When is a closed set $F \subset G$ a Carleman set? See Definition 3 in A_1. We exclude the trivial case $F = G$.

Theorem 4 (Nersesjan 1971). *Suppose F is a closed proper subset of G. Then F is a Carleman set in G if and only if F satisfies conditions (K_1), (K_2), and (A).*

A stronger, sufficient condition for F to be a Carleman set in G was given earlier by Gauthier (1969).

Proof. a) Conditions (K_1), (K_2), and (A) are necessary.

For (K_1) and (K_2) the necessity is obvious, since one obtains uniform approximation for $\epsilon(z) = \epsilon > 0$. Thus we need to show that (A) is necessary (Gauthier [1969, pp. 320–321]), and we assume (A) is not satisfied.

i) *Geometric preparations.* Suppose $\{G_n\}$ is an exhaustion of G by bounded domains G_n; that is $\overline{G}_n \subset G_{n+1}$, $\cup G_n = G$. If (A) is not satisfied, then for each $n \geqslant 2$ there exists a Jordan arc $b_n \subset F^o$ that connects a point $P_n \notin \overline{G}_n$ with a point $Q_n \in K$. Here K is a fixed, relatively compact subset of G, and we can assume $K \subset G_1$. Suppose $g_n \subset F^o$ is a Jordan domain containing the arc b_n, and let γ_n denote a subarc of b_n that connects ∂G_1 with ∂G_2.

Finally, let Γ_n denote a subarc of ∂g_n in $\overline{G}_{n+1} \setminus G_n$. Its harmonic measure with respect to g_n satisfies

$$\alpha_n(z) \geqslant \alpha_n > 0 \quad \text{for } z \in \gamma_n,$$

because γ_n is compact in g_n.

ii) *Step 1*: We prove the following interesting intermediate result:

(3.6) *If (A) is not satisfied, there exists an error function $\epsilon(z)$ defined on F and with the property: If $h \in \mathrm{Hol}(G)$ and $|h(z)| \leqslant \epsilon(z)$ $(z \in F)$, then $h = 0$ in G.*

To establish this, we choose constants c_n such that $0 < c_{n+1} < c_n$ and

$$c_n^{\alpha_n} \leqslant 1/n \qquad (n = 1, 2, \ldots),$$

and then we determine an error function $\epsilon(z)$ $(z \in F)$ such that

$$\epsilon(z) \leqslant 1 \quad \text{for } z \in F \quad \text{and} \quad \epsilon(z) \leqslant c_n \quad \text{for } z \in F \cap (\overline{G}_{n+1} \setminus G_n).$$

(Since we can assume that ∂G_n is a union of Jordan curves, this is easy to achieve.)

If now $h \in \mathrm{Hol}(G)$ and $|h(z)| \leqslant \epsilon(z)$ $(z \in F)$, and therefore in particular

$$|h(z)| \leqslant c_n \qquad (z \in \Gamma_n) \quad \text{and} \quad |h(z)| \leqslant 1 \qquad (z \in \partial g_n \subset F),$$

then the "two-constants theorem" implies

$$|h(z)| \leqslant c_n^{a_n(z)} \leqslant c_n^{\alpha_n} \leqslant 1/n \qquad (z \in \gamma_n; n = 2, 3, \ldots).$$

Here the γ_n are in the compact subset $\overline{G}_2$ of G, and because they connect ∂G_1 with ∂G_2, they have diameters $d_n \geqslant d > 0$. Hence an application of Lemma 5 immediately yields that $h = 0$ in a neighborhood of a point of $\overline{G}_2$; therefore $h = 0$ in G.

iii) *Step 2*: We show: F is not a Carleman set. Here we distinguish two cases.

Case 1: If $h \in A(F)$ and $|h(z)| < \epsilon(z)$ $(z \in F)$ with the error function from above, then $h = 0$ on F.

Case 2: There exists an $h \in A(F)$ such that $|h(z)| < \epsilon(z)$ $(z \in F)$ with the error function from above, but $h(z_0) \neq 0$ for some $z_0 \in F$.

Note: In (3.6) we require $h \in \mathrm{Hol}(G)$, not only $h \in A(F)$.

Concerning Case 1: If F were a Carleman set, then for each $f \in A(F)$ there would exist a $g \in \mathrm{Hol}(G)$ such that $|f(z) - g(z)| < \epsilon(z)$ $(z \in F)$. Since we are dealing with Case 1, we must have $f = g$ on F. But this would mean that every function $f \in A(F)$ has an analytic continuation to G. Since F is a proper subset of G, this is not possible. For then there exists a segment $s \subset G \setminus F$, and the conformal mapping f from $\hat{\mathbb{C}} \setminus s$ onto $\{w: |w| < 1\}$ belongs to $A(F)$ but is not analytic in G.

Concerning Case 2: Now we choose a new error function $\epsilon_1(z)$; suppose it satisfies $\epsilon_1(z) < \epsilon(z) - |h(z)|$ $(z \in F)$ as well as $\epsilon_1(z_0) < |h(z_0)|$. If F were a Carleman set, then for the h above there would exist a $g \in \mathrm{Hol}(G)$ such that $|h(z) - g(z)| < \epsilon_1(z)$ $(z \in F)$; that is,

$$|g(z)| \leqslant |h(z)| + |h(z) - g(z)| < |h(z)| + \epsilon_1(z) < \epsilon(z) \qquad (z \in F).$$

By (3.6) we have $g = 0$ in G, and for $z = z_0$ we find that $|h(z_0)| < \epsilon_1(z_0)$, in contradiction to the construction of $\epsilon_1(z)$.

In both cases F fails to be a Carleman set when (A) is not satisfied.

b) Conditions (K_1), (K_2), and (A) are sufficient.

Here we must refer the reader to the original work by Nersesjan. The method of proof introduced by Keldysh and Lavrentiev in 1939 and developed further by Mergelyan in 1952 is used again, and the desired function $g \in \mathrm{Hol}(G)$ is represented as $g = \Sigma R_n$, where the R_n are rational functions. Of central importance is a lemma the origin of which goes back to Lavrentiev [1936, p. 25]; similar lemmas can be found in Roth [1938].

Lemma. *Suppose F is compact in $\mathbb{C}$, that $\mathbb{C} \setminus F$ consists of finitely many components, and that G is an open subset of F with $\partial G \subset \partial F$. Then, for each $\epsilon > 0$, there exists a rational function $R(z, G, \epsilon)$ such that*

$$|R(z, G, \epsilon)| < \epsilon \qquad \text{for } z \in G \setminus (\partial G)_\epsilon,$$

$$|R(z, G, \epsilon) - 1| < \epsilon \qquad \text{for } z \in F \setminus G_\epsilon,$$

$$|R(z, G, \epsilon)| < C \qquad \text{for } z \in F.$$

Here C is an absolute constant, and N_ϵ denotes the ϵ-neighborhood of the set N.

For the proof of the lemma elements of the proof of Mergelyan's theorem are used.

Remarks about § 3

Without aspiring to completeness, we point out several works that are connected with Carleman's theorem.

1) Hoischen [1967] gives an interesting, largely constructive proof of Carleman's theorem. His work is based on the Gauss transform

$$F_\lambda(z) := \frac{\lambda}{\sqrt{\pi}} \int_a^b e^{-\lambda^2 (z-t)^2} f(t)dt \qquad (z \in \mathbb{C}; \lambda > 0)$$

of a function f continuous on $[a, b]$. The "hump" property of the kernel is important; it implies that $F_\lambda(x) \to f(x)$ $(a < x < b)$ and $F_\lambda(x) \to 0$ $(x < a, x > b)$, as $\lambda \to \infty$. In another work, Hoischen [1970] investigates the approximation of $f \in C(\mathbb{R})$ by entire Dirichlet series.

2) Numerous authors consider approximation with interpolation or approximation of a function f and its derivatives. In this context we cite works by Gauthier and Hengartner [1977], Hoischen [1975], Kaplan [1955], Nersesjan [1973], [1978], [1980], Rubel and Venkateswaran [1976], and Sinclair [1965]. We mention a result due to Nersesjan:

Suppose $G = \{z: |z| < R\}$ $(0 < R \leqslant \infty)$, and let γ_k denote countably many Jordan arcs such that each γ_k lies in $\{z: r_k < |z| < R\}$, where $r_k \to R$ $(k \to \infty)$. We set $L = \cup \gamma_k$ and require that f is n-times continuously differentiable on L. Finally, suppose $\epsilon(x)$ is continuous and positive in $(0, R)$. Then there exists a function $g \in \mathrm{Hol}(G)$ such that

$$|f^{(k)}(z) - g^{(k)}(z)| < \epsilon(|z|) \qquad (z \in L; k = 0, 1, 2, \ldots, n).$$

3) Carleman's theorem admits the following extension to $\mathbb{R}^n$ (Scheinberg [1976]). For $f \in C(\mathbb{R}^n)$ and a positive error function $\epsilon \in C(\mathbb{R}^n)$ there exists a $g \in \mathrm{Hol}(\mathbb{C}^n)$ such that

$$|f(x) - g(x)| < \epsilon(x) \quad \text{for } x \in \mathbb{R}^n.$$

Here x is the real part of $z \in \mathbb{C}^n$.

§4. Approximation with certain error functions

In this section we deal with several additional questions in connection with the special case $G = \mathbb{C}$; that is, we have approximation on a closed set $F \subset \mathbb{C}$ by entire functions g. We attain tangential approximation if F satisfies conditions (K_1), (K_2), and (A). Unfortunately, important sets F do not satisfy condition (A), for example, angular sectors and parallel strips. Hence the following questions arise:

1) For which error functions $\epsilon(z)$ is ϵ-approximation possible on F, if F satisfies (K_1) and (K_2) but not (A)? We already became acquainted with initial results in this direction in §3, B_1.

2) Can anything be said about the growth of the approximating entire function? In 1945, Keldysh took up these questions for the first time in a short but significant work. The work contains no proofs; but these are presented fully in Mergelyan's report. Later works concerning these problems are due to Arakeljan [1959, 1961, 1962, 1964] and Ter-Israeljan [1971].

The proofs of the following theorems are all technically quite complicated. Since they are carefully carried out by Mergelyan [1952, pp. 333-363] and especially by Fuchs [1968, pp. 41-73], we limit ourselves to an enumeration of the most important results.

A. ϵ-approximation without condition (A)

First, Carleman's formula (see Boas [1954, p. 2]) implies: If F denotes the half-plane $\{z\colon \operatorname{Re} z \geqslant 0\}$ and $h \in A(F)$, $h \neq 0$, then

$$\int_{-\infty}^{+\infty} \frac{\log |h(iy)|}{1+y^2}\, dy > -\infty \qquad (z = x + iy);$$

that is, along the imaginary axis h cannot tend to zero too fast. Hence, if $f \in A(F)$ allows ϵ-approximation on F by an entire function g:

$$|f(z) - g(z)| < \epsilon(z) \qquad (z \in F),$$

then the condition

$$\int_{-\infty}^{+\infty} \frac{\log \epsilon(iy)}{1+y^2}\, dy > -\infty \tag{4.1}$$

must be satisfied, except in the trivial case where f itself is entire. For example, $\epsilon(z) = e^{-|z|}$ is not an admissible error function for the set F above.

A small variation of this reasoning (see Fuchs [1968, p. 39]) shows: If

$$F = \{z\colon \operatorname{Re} \sqrt{z} \geqslant 1\}$$

(exterior of a parabola), then the condition

$$\int_{1}^{\infty} \frac{\log \epsilon(t)}{t^{3/2}}\, dt > -\infty \tag{4.2}$$

must be satisfied if $\epsilon(z) = \epsilon(|z|)$ is to be an admissible error function for this set F.

Arakeljan ([1959], [1961], and [1964a]) shows that in the general case, (4.2) is also sufficient.

Theorem 1. *Suppose the closed set $F \subset \mathbb{C}$ satisfies conditions (K_1) and (K_2) from §2, and suppose $\epsilon(t)$ is continuous and positive for $t \geqslant 0$ and satisfies*

$$\int_{1}^{\infty} t^{-3/2} \log \epsilon(t)\, dt > -\infty. \tag{4.2}$$

Then every function $f \in A(F)$ allows ϵ-approximation on F with $\epsilon(z) = \epsilon(|z|)$ $(z \in F)$. The statement does not remain valid for every F if (4.2) *is violated.*

This theorem sharpens Theorem 3 in §2; (K_1) and (K_2) guarantee not only uniform approximation, but even ϵ-approximation for certain error functions.

Examples: $\epsilon(t) = \exp(-t^{1/2})$ does not satisfy (4.2), but $\epsilon(t) = \exp(-t^{1/2-\eta})$ does for every $\eta > 0$.

The proof of Theorem 1 depends on Lemma 3 in §3. Solving a suitable Dirichlet problem in a simply connected domain $g \supset F$, we construct a function u harmonic in g such that $u(z) \leqslant \log \epsilon(|z|)$ $(z \in F)$. If v is a conjugate of u, an application of Lemma 3 to $\psi = u + iv$ yields the result.

Corollary. *If F lies in a sector*

$$W_\alpha = \{z\colon |\arg z| \leqslant \alpha/2\} \quad (0 < \alpha \leqslant 2\pi)$$

with opening α, then condition (4.2) *in Theorem* 1 *can be replaced by the weaker condition*

$$\int_1^\infty t^{-(\pi/\alpha)-1} \log \epsilon(t)\, dt > -\infty. \tag{4.2'}$$

In this case, even $\epsilon(z) = \exp(-|z|^{(\pi/\alpha)-\eta})$ is an admissible error function for every $\eta > 0$. There are corresponding theorems if F is contained in a parallel strip.

B. Growth of the approximating function

Often it is important to have some information about the growth of the approximating entire function g. However, now the approximation is measured only on a subset of F. We let the following specific result suffice.

Theorem 2. *Suppose $f \in A(W_\alpha)$ $(0 < \alpha \leqslant \pi)$ and*

$$|f(z)| \leqslant K\, e^{k|z|^\rho} \quad (z \in W_\alpha),$$

where $\rho = \pi/(2\pi - \alpha)$ (so that $\frac{1}{2} < \rho \leqslant 1$); suppose further that $\epsilon > 0$ and $\delta > 0$ are given. Then there exists an entire function g such that

$$|f(z) - g(z)| \leqslant \epsilon e^{-|z|^\rho} \quad (z \in W_{\alpha-\delta})$$

and

$$|g(z)| \leqslant C\, e^{c|z|^{\rho}} \qquad (z \in \mathbb{C})$$

for certain constants c and C.

For $\alpha = \pi$ it follows that $\rho = 1$, and the approximating function is of exponential type. Keldysh [1945b, p. 240] and Mergelyan [1952, p. 353] allow more general growth of f. Corresponding theorems hold for functions in parallel strips.

C. The special case $F = \mathbb{R}$

Suppose f is continuous on $\mathbb{R}$. If one wants a uniform approximation of f on $\mathbb{R}$ by an entire function g, that is, $|f(x) - g(x)| < \epsilon$ $(x \in \mathbb{R})$, then g will in general have strong growth in the plane, even if f is bounded on $\mathbb{R}$. For if f oscillates strongly on $\mathbb{R}$, then Re g also oscillates strongly, so that $\frac{d}{dx}$ (Re $g(x)$) must assume large values on a dense sequence of points; but this means that g' must assume large values and hence also g, at any rate in $\mathbb{C}$.

But Keldysh discovered that the growth of g can be limited, provided assumptions are made about the growth of f *and* f'. The following is a refinement of his result by Arakeljan [1963].

Theorem 3. *Suppose* $f \in C^1(\mathbb{R})$, *and set*

$$M(r) := \max\{|f(x)| : |x| \leqslant r\},$$

$$\mu(r) := \max\{|f'(x)| : |x| \leqslant r\}.$$

Then, for $\epsilon > 0$, *there exists an entire function g such that*

$$|f(x) - g(x)| < \epsilon \qquad (x \in \mathbb{R})$$

and

$$|g(z)| \leqslant \exp[A(|z| + 1) \cdot B(|z|)] \qquad (z \in \mathbb{C}),$$

where A is a constant and

$$B(|z|) = \max\left\{\frac{1}{\epsilon}\,\mu(t)\log\left[\frac{1}{\epsilon}\,c(f)\,M(t)\,\mu(t)\right] : 0 \leqslant t \leqslant 2|z| + 1\right\}.$$

Here $c(f)$ *is positive and depends only on f.*

In particular, if f and f' are bounded on $\mathbb{R}$, then g can be chosen to be of exponential type; and if $\mu(r) = O(r^{\alpha})$ $(r \to \infty)$ for some $\alpha > 0$, then g can still be chosen to be of order $\alpha + 1$, possibly of maximal type. This result also originated with Keldysh.

§5. Some applications of the approximation theorems

It was recognized early that theorems about the approximation of functions on noncompact sets allow the construction of analytic functions with prescribed boundary behavior. For example, if one chooses in Carleman's theorem the function f continuous on $\mathbf{R}$ such that $f(\mathbf{R}) = \mathbf{C}$ (Peano curve), then the theorem immediately yields an entire function g, for which $g(\mathbf{R})$ is dense in $\mathbf{C}$ (Kaplan).

In the following we present some important applications of the approximation theorems: applications to the boundary behavior of entire functions (Part A) and to the boundary behavior of functions in the unit disk (Part B). In Part C we discuss the connection between uniqueness theorems and approximation, and Part D contains various smaller contributions. We shall not take up applications in the Nevanlinna theory; but see the remarks at the end of §5.

A. Radial boundary values of entire functions

In this section, we consider the boundary behavior of entire functions f, for which

$$\lim_{r\to\infty} f(re^{i\phi}) =: F(e^{i\phi}) \tag{5.1}$$

exists for all ϕ as a finite or infinite limit; hence $f(z) = (\sin z)/z$ is allowed, but $f(z) = \sin z$ is not. The *problem* consists of characterizing the possible functions F. A. Roth called F the "radial limit function" (Strahlengrenzwertfunktion) and characterized it by its properties. Her work [1938] represents the first important example of how approximation theorems can be used to study the boundary behavior of analytic functions.

We begin with an auxiliary consideration before we investigate F. Let W and W' denote open sectors with vertex at 0, such as $W = \{z: \alpha < \arg z < \beta\}$.

Lemma 1. *For every sector W there exists a subsector W' with the following property: Either f is bounded in W', or $1/f$ is bounded in W' for $z \to \infty$.*

Proof. Suppose f is unbounded *and* $1/f$ is unbounded for $z \to \infty$ in every subsector W' of W. Then we construct a sequence $\{I_k\}$ of closed ϕ-intervals and corresponding sectors

$$W_k = \{z: \arg z \in I_k\}$$

as follows.

We choose $z_1 \in W$ such that $|f(z_1)| > 1$ and choose $I_1 \ni \arg z_1$ such that $|f(z)| > 1$ for $|z| = |z_1|$, $\arg z \in I_1$. In W_1 we find z_1' such that $|z_1'| > |z_1|$ and $|f(z_1')| < 1$; therefore $|f(z)| < 1$ for $|z| = |z_1'|$, $\arg z \in I_1' \subset I_1$.

Next we choose z_2 such that $\arg z_2 \in I_1'$, $|f(z_2)| > 2$; hence $|f(z)| > 2$ for $|z| = |z_2|$, $\arg z \in I_2 \subset I_1'$. And in W_2 we find z_2' such that $|z_2'| > |z_2|$ and $|f(z_2')| < \frac{1}{2}$; hence $|f(z)| < \frac{1}{2}$ for $|z| = |z_2'|$, $\arg z \in I_2' \subset I_2$.

And so on. If we now let $\phi_0 \in \cap I_k = \cap I_k'$ and $r_k = |z_k|$, $r_k' = |z_k'|$, then $r_k \to \infty$ and therefore $r_k' \to \infty$, and we have

$$f(r_k e^{i\phi_0}) \to \infty \quad (k \to \infty), \qquad f(r_k' e^{i\phi_0}) \to 0 \quad (k \to \infty).$$

This contradicts our overall assumption (5.1), and the lemma is proved.

Now we proceed to derive properties of the function F defined on $C := \{z\colon |z| = 1\}$.

Theorem 1. a) *On C, the function F is of Baire class* 0 *or* 1.

b) *There exists an open set* $M = \cup_{k=1}^{\infty} b_k$ *on C with the following properties:*

i) *M is everywhere dense on C*;

ii) $F(e^{i\phi}) = c_k$ *for* $e^{i\phi} \in b_k$, *where the* c_k *are constants from* $\mathbb{C} \cup \{\infty\}$;

iii) *For each closed subarc* $\beta \subset b_k$ *we have*

$$\lim_{r\to\infty} f(re^{i\phi}) = c_k, \textit{ uniformly for } e^{i\phi} \in \beta.$$

Here the b_k are the countably many components of M, in this case, open arcs on C.

Proof. a) This statement follows immediately from the representation

$$F(e^{i\phi}) = \lim_{n\to\infty} f(ne^{i\phi});$$

F is the limit function of a sequence of continuous functions.

b) Suppose W' is a sector of the type mentioned in the lemma, and let $b = W' \cap C$.

If f is bounded in W', then the functions $f_n(z) := f(nz)$ form a normal family in W', and the sequence converges on the radial segment

$$s := \{re^{i\phi_0} : 1 \leqslant r \leqslant 2, e^{i\phi_0} \in W'\}:$$

$f_n(z) \to F(e^{i\phi_0})$ $(n \to \infty; z \in s)$. By Vitali's theorem (see, for example, Bieberbach [1934, p. 168]) we thus have $f_n(z) \to F(e^{i\phi_0})$ $(n \to \infty; z \in W')$, uniformly in compact subsets of W'. But this means $f(re^{i\phi}) \to F(e^{i\phi_0})$ $(r \to \infty; e^{i\phi} \in b)$, uniformly in compact subsets of b, and in particular,

$$F(e^{i\phi}) = F(e^{i\phi_0}), \quad \text{if } e^{i\phi} \in b.$$

If $1/f$ is bounded in W' for $z \to \infty$, it follows analogously that $1/f(re^{i\phi}) \to 1/F(e^{i\phi_0})$ ($r \to \infty$; $e^{i\phi} \in b$), uniformly in compact subsets of b, and in particular, we again have that $F(e^{i\phi}) = F(e^{i\phi_0})$, if $e^{i\phi} \in b$.

Result: On each such circular arc b the function F is constant (possibly ∞).

Now suppose M is the union of all such open circular arcs b. Then M is open and dense on $\{z: |z| = 1\}$, and $M = \bigcup_{k=1}^{\infty} b_k$, where the b_k are countably many *disjoint* open circular arcs.

Then F is constant on each circular arc b_k.

The last statement holds, because each closed subarc $\beta \subset b_k$ can be covered by finitely many open arcs b, on each of which F is constant. Hence $F(e^{i\phi}) = c_k$ for $e^{i\phi} \in b_k$. And by our statement above concerning uniformity, we see that

$$f(re^{i\phi}) \to c_k \ (r \to \infty), \text{ uniformly for } e^{i\phi} \in \beta \subset b_k.$$

This completes the proof of Theorem 1.

We now show that the properties cited in Theorem 1 are characteristic for F.

Theorem 2. *Suppose an open set $M = \bigcup_{k=1}^{\infty} b_k$ on C is given, where the b_k are disjoint, open subarcs of C, and suppose M is everywhere dense on C. Suppose further that F is of Baire class 0 or 1 on C and that*

$$F(e^{i\phi}) = c_k \quad \textit{for } e^{i\phi} \in b_k.$$

Then there exists an entire function g with the properties:

i) $\lim_{r\to\infty} g(re^{i\phi}) = F(e^{i\phi})$ *for each* ϕ;

ii) *This holds uniformly on every closed subarc* $\beta \subset b_k$.

Proof. a) First we produce a function h *continuous* on $\mathbb{C}$ and with property i). For this we use that F is at most of Baire class 1; that is, $F(e^{i\phi}) = \lim_{n\to\infty} h_n(e^{i\phi})$, where the h_n are continuous functions. Now we define

$$h \text{ on } |z| = n \quad \text{by} \quad h(ne^{i\phi}) := h_n(e^{i\phi}) \qquad (n = 1, 2, \ldots),$$

and we use linear interpolation between the points $ne^{i\phi}$ and $(n + 1)e^{i\phi}$ ($n = 0, 1, 2, \ldots$); we set $h(0)$ equal to 0. This function h is defined and continuous in $\mathbb{C}$, and we have

$$\lim_{r\to\infty} h(re^{i\phi}) = \lim_{n\to\infty} h(ne^{i\phi}) = \lim_{n\to\infty} h_n(e^{i\phi})$$
$$= F(e^{i\phi});$$

hence h has the correct radial limits.

b) To apply an approximation theorem, we need a closed set $F \subset \mathbb{C}$ and a function $f \in A(F)$.

To define F, we consider the *closed* sectors W_k (sketch) and construct

$$W := \cup_{k=1}^{\infty} W_k.$$

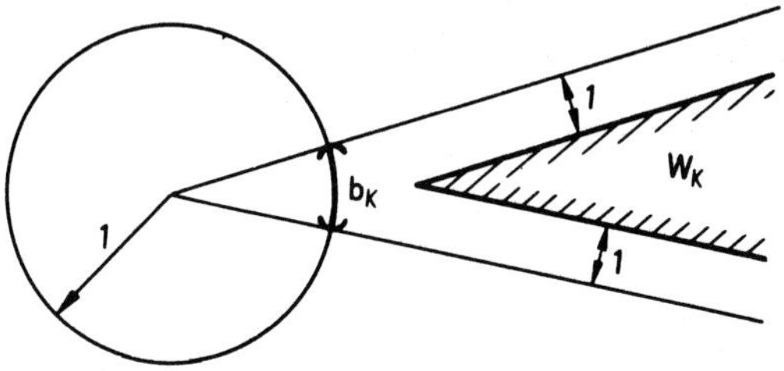

Note that W is closed in $\mathbb{C}$; for if there are infinitely many W_k, they are pushed to ∞. Next we combine the rays leading to points $\notin M$:

$$S := \{re^{i\phi} : r \geqslant 0, e^{i\phi} \notin M\};$$

S is also closed in $\mathbb{C}$, because M is open in C. Finally, we let $F := W \cup S$. This set is closed in $\mathbb{C}$, and in addition it satisfies conditions (K_1) and (K_2) for a Weierstrass set (§2).

On F we now define

$$f = \begin{cases} h & \text{on } S, \\ c_k & \text{on } W_k, \quad \text{if } c_k \neq \infty, \\ z & \text{on } W_k, \quad \text{if } c_k = \infty. \end{cases}$$

This function is continuous on F and analytic in F°, because S has no interior points; recall that M is dense on C.

By Arakeljan's approximation theorem and Theorem 2′ in §3, B, there exists an entire function g such that

$$|f(z) - g(z)| \leqslant 1/|z| \quad \text{for } z \in F. \tag{5.2}$$

This function g has the desired properties i) and ii): For if $e^{i\phi} \notin M$, that is, $re^{i\phi} \in S$, then (5.2) implies

$$f(re^{i\phi}) - g(re^{i\phi}) \to 0, \text{ so that } h(re^{i\phi}) - g(re^{i\phi}) \to 0 \qquad (r \to \infty).$$

The last statement shows that

$$\lim_{r\to\infty} g(re^{i\phi}) = \lim_{r\to\infty} h(re^{i\phi}) = F(e^{i\phi}).$$

If $e^{i\phi} \in M$, so that $e^{i\phi} \in b_k$ for some $k \in \mathbf{N}$, then $re^{i\phi} \in W_k$ for sufficiently large r; consequently

$$g(re^{i\phi}) - c_k \to 0 \quad \text{or} \quad g(re^{i\phi}) - re^{i\phi} \to 0 \qquad (r \to \infty).$$

Thus i) holds for all ϕ.

Finally, if $e^{i\phi} \in \beta \subset b_k$, then $re^{i\phi} \in W_k$ for $r \geqslant r_0$, independent of ϕ; by (5.2) we thus have

$$|g(re^{i\phi}) - c_k| \leqslant 1/r \quad \text{or} \quad |g(re^{i\phi}) - re^{i\phi}| \leqslant 1/r \qquad (r \geqslant r_0).$$

Hence ii) also holds, and this completes the proof of Theorem 2.

Remarks. 1) Of the prescribed entire function we basically required that the limit (5.1) should exist for all ϕ. If it only exists for ϕ in some interval (as is the case for $f(z) = \sin z$), then theorems analogous to Theorems 1 and 2 hold in this interval.

2) We point out the following detail. The set M in Theorem 1 is not uniquely determined by f; for example, removal of a point results in a new set, for which the theorem holds. In the proof, however, M was constructed in a unique way as the set $\cup b$, with well defined arcs b. We denote this set $\cup b$ by $M(f)$.

An easy modification of the construction above now yields a result sharper than Theorem 2, namely an entire function g for which Theorem 2 holds and for which, in addition, $M(g) = M$, where M is the prescribed set; see Roth [1938, p. 119]. This result is interesting, because $M(g)$ is closely connected with the Julia directions of g; see Section D_4.

B. Boundary behavior of functions analytic in the unit disk

Arakeljan's approximation theorems are also eminently suited for the construction of functions that are analytic in the unit disk $\mathbb{D}$ and exhibit a certain boundary behavior.

B_1. A general approximation theorem

Theorem 3. *Suppose we have countably many sets E_n on $\{z: |z| = 1\}$ ($n = 1, 2, \ldots$), where each E_n is closed and nowhere dense, and a function f continuous in $\mathbb{D}$. Then there exists a function g analytic in $\mathbb{D}$ such that*

$$g(re^{i\phi}) - f(re^{i\phi}) \to 0 \quad \text{as } r \to 1^- \tag{5.3}$$

for each $e^{i\phi} \in E = \cup E_n$.

Remarks. The theorem says that on the radii leading to E, each function continuous in $\mathbb{D}$ can be imitated by a function $g \in \mathrm{Hol}(\mathbb{D})$. – A set E of the form mentioned is called an F_σ-set of first category. If we choose for E_n Cantor sets of measure $2\pi - 1/n$, then E has measure 2π. *Statement* (5.3) *then holds* *for almost all* $\phi \in (0, 2\pi)$. – Bagemihl and Seidel [1954, p. 187] state this theorem for more general systems of "monotone boundary arcs"; the following proof can be adapted for this. (For analogous theorems in more general domains see Bagemihl and Seidel [1955].)

Proof. We apply Arakeljan's approximation theorems to the following situation.

Suppose $G = \mathbb{D}$ and $F = \bigcup_{n=1}^{\infty} F_n$, where $F_n := \{re^{i\phi}: 1 - \dfrac{1}{n+1} \leqslant r < 1,$ $e^{i\phi} \in E_n\}$ $(n = 1, 2, \ldots)$. Each set F_n is closed in $\mathbb{D}$ with $F_n^{\circ} = \phi$, and their union F is also closed in $\mathbb{D}$ with $F^{\circ} = \phi$.

Theorem 3 in §3, B_2 implies that F is a Carleman set in $\mathbb{D}$ provided F is a Weierstrass set in $\mathbb{D}$. According to Theorem 3 in §2, C_2, we thus need to verify that $\mathbb{D}^* \setminus F$ is connected, and that it is locally connected at the ideal point $\infty \in \mathbb{D}^*$. The first condition is satisfied, because $\mathbb{D} \setminus F$ is connected and has an accumulation point on $\{z: |z| = 1\}$. The second condition is satisfied, because for every point $z_0 \in \mathbb{D} \setminus F$ one can find a path $\zeta = \zeta(t)$, $t \in [0, 1)$, such that $\zeta(0) = z_0$ and $|\zeta(t)| \uparrow 1$ as $t \uparrow 1$. Here one uses that the E_n are nowhere dense on $\{z: |z| = 1\}$.

Since F has no interior points, it follows that $f \in A(F)$, and thus, for every error function $\epsilon(r)$, there exists a function $g \in \mathrm{Hol}(\mathbb{D})$ such that

$$|f(z) - g(z)| \leqslant \epsilon(|z|) \qquad (z \in F). \tag{5.4}$$

But each radius leading to a point of E is in F from some point on; this implies (5.3).

Closely connected with Theorem 3 is the question of the relationship between the uniform boundedness of an analytic function g on a set of radii and the existence of radial limits of g.

Theorem 4. *Suppose M is a set of radii in $\mathbb{D}$. In order that there exists a function $g \in \mathrm{Hol}(\mathbb{D})$ that is uniformly bounded on the radii of M but does not have a limit on any radius of $\mathbb{D}$ it is necessary and sufficient that M is nowhere dense.*

Remark. If one chooses M such that the set of end points of the radii of M is nowhere dense on $\partial\mathbb{D}$ and has measure $2\pi - \delta$ $(\delta > 0)$, then g can be uniformly bounded on the radii of M without possessing a radial limit anywhere. For $\delta = 0$ this is of course not possible, since in this case g is bounded in $\mathbb{D}$ and therefore, by Fatou's theorem, possesses radial limits almost everywhere.

Proof. The condition is necessary. For if M is not nowhere dense, then M must be dense in some sector S of $\mathbb{D}$. Every function g that is uniformly bounded on M is therefore bounded in S and consequently has limits along the radii in S.

Conversely, suppose $M \neq \phi$ is a nowhere dense set of radii in $\mathbf{D}$. We construct a function $g \in \mathrm{Hol}(\mathbf{D})$ that is uniformly bounded on the radii of M but has no radial limits. We can assume that M is closed, for otherwise we would consider $\overline{M}$.

Here again Arakeljan is helpful. We let

$$F_n := \{z = re^{i\phi} : r = 1 - 1/n; \mathrm{dist}\,(z, M) \geqslant 1/n\} \quad (n = 1, 2, \ldots);$$

these are closed sets of circular arcs on $\{z : |z| = 1 - 1/n\}$. Then we consider

$$F := M \cup (\cup_{n=1}^{\infty} F_n).$$

This set is closed in $\mathbf{D}$ and without interior points, and it satisfies the conditions for a Weierstrass set; the latter can be checked as above. It follows that $f(z) = \cos\,(\pi/(1 - |z|)) \in A(F)$ can be approximated by a $g \in \mathrm{Hol}(\mathbf{D})$ with the error function $\epsilon(r)$:

$$|f(z) - g(z)| \leqslant \epsilon(|z|) \qquad (z \in F).$$

The function g thus has no limits on the radii of M. But by the construction of F_n, every other radius meets F_n for $n > n_0$, and hence we have on these radii

$$f(re^{i\phi}) - g(re^{i\phi}) \to 0 \quad \text{for } r = 1 - 1/n, \qquad n \to \infty.$$

It follows that

$$g((1 - \frac{1}{n})\, e^{i\phi}) = (-1)^n + o(1) \qquad (n \to \infty).$$

Result: The function g does not have a limit on any radius of $\mathbf{D}$, even though g is uniformly bounded on F and therefore on the radii of M.

B_2. The Dirichlet problem for radial limits

We use the approximation theorem of Bagemihl and Seidel to prove the existence of a solution to the general Dirichlet problem for radial limits.

Theorem 5. *Suppose u and v are two real-valued, measurable functions in* $(0, 2\pi)$; *the values* $\pm\infty$ *are allowed. Then there exists a function g analytic in* $\mathbf{D}$ *with the property that*

$$\lim_{r\to 1^-} \; g(re^{i\phi}) = u(\phi) + iv(\phi) \tag{5.5}$$

for almost all $\phi \in (0, 2\pi)$.

Remarks. This theorem was proved simultaneously by Bagemihl and Seidel [1955] and byLehto [1955]. In addition, Lehto represents g by an integral of Poisson type, provided the boundary values are finite almost everywhere. For the case of infinite boundary values, a direct, but laborious construction is known (Privalov [1956, p. 225]).– The real part of g solves the analogous Dirichlet problem for harmonic functions, if the radial limits u are given as a measurable function (Kaplan [1955, p. 49]).– Finally, it is noteworthy that one can require the function g in Theorem 5 to satisfy $|g(z)| \leqslant \mu(|z|)$, where μ is a function tending to ∞ arbitrarily slowly (Kegejan [1966]).

Proof. *Step 1*: Construction of a function f *continuous* in $\mathbf{D}$ and with radial limits $u + iv$. It is enough to find a function f continuous in $\mathbf{D}$ such that $\lim_{r\to 1^-} f(re^{i\phi}) = u(\phi)$ for almost all ϕ. If $u(\phi) = \infty$ a.e., we choose $f(z) = 1/(1 - |z|)$, and if $u(\phi) = -\infty$ a.e., we choose $f(z) = -1/(1 - |z|)$. Hence we can assume that $u < +\infty$ and $u > -\infty$ on two sets of positive measure. For the bounded, measurable function $U(\phi) = \arctan u(\phi)$ we find the corresponding Poisson integral $\Phi(z)$, which, by our assumption, satisfies $|\Phi(z)| < \pi/2$ $(z \in \mathbf{D})$. It has the boundary values $\lim_{r\to 1^-} \Phi(re^{i\phi}) = U(\phi)$ a.e.; consequently the function $f(z) := \tan \Phi(z)$ is continuous in $\mathbf{D}$ and has radial limits $\tan U(\phi) = u(\phi)$ for almost all $\phi \in (0, 2\pi)$.

Step 2: Construction of a function g *analytic* in $\mathbf{D}$ and with radial limits $u + iv$. We take an arbitrary F_σ-set E of first category on $\{z: |z| = 1\}$: $E = \cup E_n$, where the E_n are closed and nowhere dense on $\{z: |z| = 1\}$. In addition, we assume E has measure 2π. Theorem 3 now yields the desired function $g \in \mathrm{Hol}(\mathbf{D})$.

Of course, our boundary value problem has many solutions. For Theorem 3 (or the somewhat stronger statement (5.4)) yields arbitrarily many different functions $g \in \mathrm{Hol}(\mathbf{D})$ with $\lim_{r\to 1^-} g(re^{i\phi}) = 0$ for almost all ϕ.

C. Approximation and uniqueness theorems

Suppose G is an arbitrary domain, F is closed in G with $F \neq G$, and $\epsilon(z)$ is a continuous and positive function defined on F. We ask when the following uniqueness theorem holds:

(5.6) $$h \in \mathrm{Hol}(G),\ |h(z)| \leqslant \epsilon(z) \quad (z \in F) \ \text{ implies } h = 0.$$

Intuitively one would say that (5.6) holds if F is a large set and the values of ϵ are small. But then approximation on F with the error function ϵ would hardly be possible; that is, if (5.6) holds, then

For every function $f \in A(F)$ there exists a $g \in \mathrm{Hol}(G)$ such that

(5.7) $$|f(z) - g(z)| < \epsilon(z) \quad (z \in F)$$

would be difficult to satisfy, and vice versa. More precisely, the following holds.

If F does not satisfy condition (A) from §3, C_1, *then there exists a function* $\epsilon(z)$, *for which* (5.6) *holds.*

We have already proved this in (3.6). In the other direction we now prove the following.

If (5.6) *holds for a function* $\epsilon(z)$, *then* (5.7) *is false for this function.*

Proof. We show: If (5.7) holds, then (5.6) fails. If F has no accumulation point in G, we simply choose a function $h \in \mathrm{Hol}(G)$ that vanishes exactly on F; our assertion holds. We can therefore assume that F has an accumulation point in G. We choose $a \in G \setminus F$, $f = (z - a)^{-1} \in A(F)$, and $g \in \mathrm{Hol}(G)$ such that $|f(z) - g(z)| < \frac{1}{2}\epsilon(z)$ $(z \in F)$. Here

$$f(b) - g(b) \neq 0 \quad \text{for some } b \in F,$$

since otherwise we would have $f(z) = g(z)$ on F, and therefore on G. Hence $|f(b) - g(b)| > \epsilon(b)/N$ for some natural number $N > 1$. Now we approximate f within $\epsilon(z)/N$; that is,

$$|f(z) - h(z)| < \epsilon(z)/N \qquad (z \in F) \quad \text{for some } h \in \mathrm{Hol}(G).$$

Then (5.6) fails for the function $g - h$: We do have $g - h \in \mathrm{Hol}(G)$ and $|g(z) - h(z)| < \epsilon(z)$ $(z \in F)$, but

$$|g(b) - h(b)| \geqslant |g(b) - f(b)| - |f(b) - h(b)| > 0.$$

This establishes our assertion.

In the case where $G = \mathbf{D} = \{z\colon |z| < 1\}$, the connection between approximation and statements about uniqueness can be expressed more clearly. We call a closed set $F \subset \mathbf{D}$ an *asymptotic uniqueness set* if $h \in \mathrm{Hol}(\mathbf{D})$, $h(z) \to 0$ for $|z| \to 1$ in F implies $h = 0$. And we call a closed set $F \subset \mathbf{D}$ a *set of asymptotic approximation* if:

For every function $f \in A(F)$ there exists a $g \in \mathrm{Hol}(\mathbf{D})$ such that $|f(z) - g(z)| \to 0$ for $|z| \to 1$ in F.

Theorem 6. *The closed set* $F \subset \mathbf{D}$ *is an asymptotic uniqueness set if and only if F is not a subset of a set* $(\neq \mathbf{D})$ *of asymptotic approximation.*

Compare this with the work by Brown, Gauthier, and Seidel [1975, p. 6], where results concerning the approximation by meromorphic functions can also be found. Stray [1978] deals with asymptotic approximation on more general open sets in place of $\mathbf{D}$.

D. Various further constructions

D_1. Prescribed boundary behavior along countably many curves

No doubt it has been noticed that in Theorem 3, for the construction of a function analytic in $\mathbb{D}$ and with prescribed boundary behavior, the approximated function f was assumed to be *continuous* in $\mathbb{D}$. If one desires a certain boundary behavior only along *countably many* curves, one can prove more.

Theorem 7. *Suppose γ_n ($n \in \mathbb{N}$) are countably many disjoint boundary paths in $\mathbb{D} = \{z\colon |z| < 1\}$; that is,*

$$\gamma_n : z = z_n(t) \text{ continuous and bijective for } 0 \leqslant t < 1, \text{ with } |z_n(t)| \to 1 \text{ for } t \to 1.$$

Suppose the function f is defined on $\cup\gamma_n$ and continuous on γ_n for each n. Then there exists a function g analytic in $\mathbb{D}$ such that

$$g(z) - f(z) \to 0 \quad \text{for } |z| \to 1 \text{ on } \gamma_n \tag{5.8}$$

for each $n \in \mathbb{N}$.

This theorem can be applied, for example, if all γ_n are spirals that wind toward $\partial\mathbf{D}$, and if one sets $f(z) = c_n$ for $z \in \gamma_n$.

Proof. The path γ_n either is entirely contained in $\{z\colon 1 - 1/(n+1) \leqslant |z| < 1\}$, or it has a last point of intersection $z_n(t_n)$ with $\{z\colon |z| = 1 - 1/(n+1)\}$. In the first case, we let $\tilde{\gamma}_n = \gamma_n$, in the second case $\tilde{\gamma}_n = \{z = z_n(t)\colon t_n \leqslant t < 1\}$; the $\tilde{\gamma}_n$ are the terminal segments of γ_n. We let

$$F := \cup_{n=1}^{\infty} \tilde{\gamma}_n;$$

clearly F is closed in $\mathbb{D}$, also $F^{\circ} = \phi$, and f is continuous on F, so that $f \in A(F)$.

We now check conditions (K_1) and (K_2) of Arakeljan's theorem. Clearly (K_1) holds, because $\mathbb{D} \setminus F$ is connected. To show that (K_2) holds, assume r $(0 < r < 1)$ is arbitrary. Only finitely many $\tilde{\gamma}_n$ meet $\{z\colon |z| = r\}$, and suppose their last intersections occur for the values τ_n of the parameter. The initial segments $\gamma_n^* = \{z = z_n(t)\colon 0 \leqslant t \leqslant \tau_n\}$ for these n lie in a compact subset of $\mathbb{D}$; hence there exists a number R $(r < R < 1)$ such that $\{z\colon |z| = R\} \cap \gamma_n^* = \phi$. If we now choose $z \in (\mathbb{D} \setminus F) \cap \{z\colon R < |z| < 1\}$ and move toward $\partial\mathbf{D}$ (radially, say), we will meet a terminal path $\tilde{\gamma}_n \setminus \gamma_n^*$ or one of the $\tilde{\gamma}_n$ that are entirely contained in $\{z\colon |z| > r\}$. In both cases one can travel along this path in $\mathbb{D} \setminus F$ toward $\partial\mathbf{D}$ without leaving $\{z\colon r < |z| < 1\}$. Hence $\mathbf{D}^* \setminus F$ is locally connected at the ideal point ∞.

By Arakeljan's theorems (Theorems 3 in §2, C_2 and §3, B_2), there exists a function g analytic in $\mathbb{D}$ such that $g(z) - f(z) \to 0$ ($|z| \to 1$ in F); this implies (5.8).

The theorem and proof remain valid if $\mathbb{D}$ is replaced by an arbitrary domain G. The verification of (K_2) is a bit more troublesome. See Gauthier and Seidel [1971, p. 461] and Kaplan [1955, p. 44].

D_2. Analytic functions with prescribed cluster sets

Theorem 7 can be used to construct functions analytic in $\mathbb{D}$ that have prescribed cluster sets along countably many, disjoint boundary paths. If

$$\gamma\colon z = z(t) \text{ continuous and bijective for } 0 \leqslant t < 1$$

is an arbitrary Jordan arc (open on the right), and if f is defined on γ, then the *cluster set* of f on γ is the set of all limit points of sequences $\{f(z(t_n))\}$, $t_n \to 1$. We denote it by $C(f, \gamma)$; it is a subset of $\hat{\mathbb{C}} = \mathbb{C} \cup \{\infty\}$.

Clearly, $C(f, \gamma)$ is always closed in $\hat{\mathbb{C}}$; and if f is continuous on γ, then $C(f, \gamma)$ is even connected, hence a continuum in $\hat{\mathbb{C}}$. We need a converse:

Lemma 2. *If K is a continuum in $\hat{\mathbb{C}}$, then there exists a function f continuous on γ whose cluster set on γ equals K: $C(f, \gamma) = K$.*

Proof. We may assume that $\gamma = \{t\colon 0 \leqslant t < 1\}$. If K is compact in $\mathbb{C}$, we cover K with finitely many open disks $U_j^{(n)}$ with radii $1/n$, and we form $G_n = \cup_j U_j^{(n)}$. If only disks with $K \cap U_j^{(n)} \neq \phi$ are used, then G_n itself is connected, and therefore a domain. We find a polygonal path $P_n \subset G_n$ that begins and terminates at a fixed point of K and that meets each disk $U_j^{(n)}$. Finally, let f_n denote a continuous mapping of the interval $I_n := [1 - 1/n, 1 - 1/(n+1)]$ onto P_n.

If one carries out this construction for $n = 1, 2, \ldots$ and then defines f by

$$f|_{I_n} = f_n,$$

then f is continuous on $[0, 1)$ and has K as its cluster set on $[0, 1)$.

If $\infty \in K$, the proof needs minor modifications.

As an application of Theorem 7 we now prove the following result.

Theorem 8. *Suppose γ_n $(n \in \mathbb{N})$ are countably many disjoint boundary paths in $\mathbb{D}$, and suppose K_n are countably many continua in $\hat{\mathbb{C}}$. Then there exists a function g analytic in $\mathbb{D}$ that has K_n as its cluster set on γ_n for each $n \in \mathbb{N}$.*

This statement is somewhat more general than that of Bagemihl and Seidel [1954, p. 194]; there the γ_n are assumed to be *monotone* boundary paths.

Proof. On each boundary path γ_n we define a continuous function f_n that has K_n as its cluster set (Lemma 2); we then define f on $\cup\gamma_n$ by $f|\gamma_n = f_n$ and apply Theorem 7. Now (5.8) implies that g and f have the same cluster set K_n on γ_n.

D_3. Schneider's noodles

Suppose γ_1 and γ_2 are two Jordan arcs in $\mathbb{C}$:

$$\gamma_{1,2}: z = z_{1,2}(t) \text{ continuous and bijective for } 0 \leqslant t < 1,$$

where $z_1(0) = z_2(0) = 0$ and $|z_{1,2}(t)| \to \infty$ for $t \to 1$. The two arcs are to have only the origin in common. Then the curve $\gamma_1 \cup \gamma_2$ decomposes $\mathbb{C}$ into two subdomains G_1 and G_2, which can, for example, be "noodles" if γ_1 and γ_2 are spirals.

Then there exists an entire function g that is bounded in G_1 but unbounded in G_2.

To prove this, suppose Γ is an unbounded Jordan arc in G_2 and set $F = \overline{G}_1 \cup \Gamma$, which is closed in $\mathbb{C}$. Further, let

$$f(z) = \begin{cases} 0 & \text{for } z \in \overline{G}_1, \\ z & \text{for } z \in \Gamma, \end{cases}$$

so that $f \in A(F)$. Now one shows as in the proof of Theorem 7 (but even more easily) that the conditions of Arakeljan's theorem are satisfied. Hence there exists an entire function g such that $|f(z) - g(z)| \leqslant 1$ for $z \in F$. This implies our assertion.

D_4. Julia directions of entire functions

Here we deal with a question concerning the value distribution of entire functions.

Definition. *We call ϕ a Julia direction of the entire function f, if in every sector* $\{z: |\arg z - \phi| < \delta\}$, $\delta > 0$, *the function f assumes each complex value infinitely often, with at most one exception.*

Example. The function $f(z) = e^z$ has the Julia directions $\pm\pi/2$, and in each case the Julia exceptional value is 0.

The *problem* is to describe the set

$$J(f) = \{e^{i\phi}: \phi \text{ is a Julia direction of } f\},$$

which is always closed, or to discuss the exceptional values.

a) For the class of entire functions f, which we dealt with in Part A of this section and for which $\lim_{r\to\infty} f(re^{i\phi})$ exists for all ϕ, the set $J(f)$ can be described as follows.

With the notation of Part A we have

$$J(f) = C \setminus M(f).$$

Proof. We show that $M(f) = C \setminus J(f)$. If $e^{i\phi} \in M(f) = \cup b$, then there exists a sector W' of the type mentioned in Lemma 1 such that $e^{i\phi} \in W'$. Either f is bounded in W', or $1/f$ is bounded in W' for $z \to \infty$. In either case ϕ cannot be a Julia direction.

Conversely, if $e^{i\phi} \notin J(f)$, then there exists a sector $W \ni e^{i\phi}$ in which f assumes *two* values at most finitely often. Thus the functions $f_n(z) := f(nz)$ form a normal family in W, and $\{f_n(z)\}$ converges for all $z \in W$, and therefore it converges uniformly on compact subsets of W. (Here $f_n \to \infty$ is allowed.) But then each subsector W' of W has the property mentioned in Lemma 1, and therefore $e^{i\phi} \in M(f)$.

Corollary. *If f belongs to the class of functions from Part A, then $J(f)$ is closed and nowhere dense on* $\{z\colon |z| = 1\}$, *and each such set is a $J(f)$ for some suitable entire function f.*

b) $J(f)$ has also been characterized for other classes of entire functions. For every closed set $J \neq \phi$ on $\{z\colon |z| = 1\}$ there exists an entire function f of finite order such that $J(f) = J$ (Anderson and Clunie [1969]). The special case $J = \{z\colon |z| = 1\}$ was discussed already by Julia and again by Cain [1974]. Drasin and Weitsman [1976] have characterized $J(f)$ for entire functions of order ρ, $0 < \rho < \infty$.

c) Answering a question of C. Rényi, Barth and Schneider [1972] have constructed an entire function f, for which $\phi = 0$ and $\phi = \pi$ are Julia directions with *different* exceptional values 1 and 0. The construction uses Theorem 1 in §4 in the weaker form, already proved by Keldysh, in which the error function $\epsilon(t) = \exp(-t^{1/2-\eta})$ $(\eta > 0)$ occurs.

Remarks about §5

1. Further applications of approximation theorems to the construction of entire functions and functions analytic in $\mathbf{D}$ that have a particular boundary behavior were given by Barth and Schneider; see the paper by Schneider [1980] in the Durham Conference Proceedings. We mention three such constructions.

a. There is a function $f \in \mathrm{Hol}(\mathbf{D})$ such that to each $\phi \in [0, 2\pi]$ there exists a path $P_\phi \subset \mathbf{D}$, tangent to the radius from 0 to $e^{i\phi}$, for which

$$f(z) \to 0 \quad \text{as } z \to e^{i\phi} \quad \text{along } P_\phi.$$

b. Given any function $p(r)$ $(0 \leqslant r < 1)$ with $0 < p(r) \uparrow \infty$ as $r \to 1$, there exist f and g in $\mathrm{Hol}(\mathbf{D})$ such that $|f(z)| \leqslant p(r)$ on $|z| = r$, $|g(z)| \leqslant p(r)$ on $|z| = r$, and such that f possesses no radial limits, whereas g has radial limits 0 at almost all points of $\partial\mathbf{D}$ without vanishing identically.

These examples are important, since they serve to illustrate the precision of the classical theorems of Lusin and Privalov and of Fatou and Riesz concerning the boundary behavior of functions in Hol($\mathbb{D}$).

2. Arakeljan's theorem about tangential approximation by entire functions has attained importance in connection with the so-called *inverse problem of Nevanlinna theory*; see Wittich [1968, Chapter 8]. Suppose the function f is meromorphic in $\mathbb{C}$, and let $\delta(a) = \delta(a, f)$ $(a \in \hat{\mathbb{C}})$ denote the Nevanlinna deficiency of f at a. It is well known that $\Sigma_{a\in\hat{\mathbb{C}}}\,\delta(a) \leqslant 2$. If f is entire, we have instead that $\Sigma_{a\in\mathbb{C}}\,\delta(a) \leqslant 1$, because $\delta(\infty) = 1$.

The question is whether, for a given function $\delta(a)$ that satisfies these conditions, there exists a meromorphic or entire function f such that $\delta(a, f) = \delta(a)$. For a long time it was not known whether functions f exist at all for which $\delta(a, f) > 0$ for infinitely many values of a. Goldberg (1954, 1959) was the first to construct such functions; Fuchs and Hayman [1962] (see also Hayman [1964, Chapter 4.1]) produced an entire function f with prescribed deficiencies $\delta(a_k, f) > 0$ at countably many prescribed points $a_k \in \mathbb{C}$.

In 1966 Arakeljan showed: *For each $\rho > 1/2$ and each sequence $\{a_k\}$ of complex numbers there exists an entire function f of order ρ such that* $\delta(a_k, f) > 0$.

A proof can be found in Fuchs [1968, Chapter 8]. Meanwhile, Drasin ([1974a], [1974b]) has completely solved the entire inverse problem of Nevanlinna theory, incorporating also the index of multiplicity $\theta(a) = \theta(a, f)$ and without using approximation theory. Instead, he uses essential tools from the theory of quasiconformal mappings.

REFERENCES

Books or survey articles concerning the topic of this book are marked with *. We use MR, Zbl, JB to refer to reviews in Mathematical Reviews, the Zentralblatt für Mathematik, or the Jahrbuch über die Fortschritte der Mathematik. In each case the *page* numbers are given. By *I*:2, for example, we indicate that a work is referred to in Chapter I, §2.

D. AHARONOV and J. L. WALSH, 1971: Some examples in degree of approximation by rational functions. Trans. Amer. Math. Soc. *159*, 427–444. MR *44*, 1268. *I*:4.

S. YA. AL'PER, 1955: On uniform approximations of functions of a complex variable in a closed region (Russian). Izv. Akad. Nauk SSSR Ser. Mat. *19*, 423–444. MR *17*, 729. *I*:6.

S. YA. AL'PER and G. I. KALINOGORSKAJA, 1969: The convergence of Lagrange interpolation polynomials in the complex domain (Russian). Izv. Vysš. Učebn. Zaved. Matematika *1969*, no. 11 (90), 13–23. MR *41*, 700. *II*:2.

J. M. ANDERSON and J. CLUNIE, 1969: Entire functions of finite order and lines of Julia. Math. Z. *112*, 59–73. MR *40*, 822. *IV*:5.

J.-E. ANDERSSON, 1975: On the degree of polynomial and rational approximation of holomorphic functions. Dissertation. Göteborg. *I*:6.

______, 1976: On the degree of weighted polynomial approximation of holomorphic functions. Anal. Math. *2*, 163–171. MR *56*, 1633. *I*:6.

M. I. ANDRAŠKO, 1964: Approximation in the mean of analytic functions in domains with a smooth boundary (Russian). Problems Math. Phys. and Theory of Functions, 3–11; Izdat. Akad. Nauk Ukrain. SSR, Kiev. MR *35*, 1042. *I*:6.

N. C. ANKENY and T. J. RIVLIN, 1955: On a theorem of S. Bernstein. Pacific J. Math. *5*, 849–852. MR *17*, 833. *I*:4.

N. U. ARAKELJAN, 1959: Refinement of some Keldyš theorems on asymptotic approximation by entire functions (Russian). Dokl. Akad. Nauk SSSR *125*, 695–698. MR *21*, 927. *IV*:4.

______, 1961: Asymptotic approximation by entire functions in infinite regions (Russian). Mat. Sb. (N.S.) *53 (95)*, 515–538. Translation: Translations Amer. Math. Soc. *43* (1964), 169–193. MR *24 A*, 154. *IV*:4.

______, 1962: Uniform approximation by entire functions on unbounded continua and an estimate of the rate of their growth (Russian). Akad. Nauk Armjan. SSR Dokl. *34*, 145–149. MR *25*, 432. *IV*:4.

______, 1963: Uniform approximation by entire functions with an estimate of their growth (Russian). Sibirsk. Mat. Ž. *4*, 977–999. MR *28*, 49. *IV*:4.

______, 1964a: Uniform and asymptotic approximation by entire functions on unbounded closed sets (Russian). Dokl. Akad. Nauk SSSR *157*, 9–11. Translation: Soviet Math. Dokl. *5* (1964), 849–851. MR *29*, 465. *IV*:2,4.

______, 1964b: Uniform approximation on closed sets by entire functions (Russian). Izv. Akad. Nauk SSSR Ser. Mat. *28*, 1187–1206. MR *30*, 52. *IV*:3.

______, 1966: Entire functions of finite order with an infinite set of deficient values (Russian). Dokl. Akad. Nauk SSSR *170*, 999–1002. Translation: Soviet Math. Dokl. *7* (1966), 1303–1306. MR *34*, 1114. *IV*:5.

______, 1968: Uniform and tangential approximations by analytic functions (Russian). Izv. Akad. Nauk Armjan. SSR Ser. Mat. *3*, 273–286. MR *43*, 104. *IV*:2,3.

______, 1970: Entire and analytic functions of bounded growth with an infinite set of deficient values (Russian). Izv. Akad. Nauk Armjan. SSR Ser. Mat. *5*, 486–506. MR *44*, 347. *IV*:5.

______, *1971: Certain questions of approximation theory and the theory of entire functions (Russian). Mat. Zametki *9*, 467–475. Translation: Math. Notes *9* (1971), 267–271. MR *44*, 90.

N. U. ARAKELJAN and V. A. MARTIROSJAN, 1977: Uniform approximations in the complex plane by polynomials with gaps (Russian). Dokl. Akad. Nauk SSSR *235*, 249–252. Translation: Soviet Math. Dokl. *18* (1977), 901–904. MR *56*, 1199. *III*:2.

F. BAGEMIHL and W. SEIDEL, 1954: Some boundary properties of analytic functions. Math. Z. *61*, 186–199. MR *16*, 460. *IV*:5.

______, 1955: Regular functions with prescribed measurable boundary values almost everywhere. Proc. Nat. Acad. Sci. U.S.A. *41*, 740–743. MR *17*, 249. *IV*:5.

K. F. BARTH and W. J. SCHNEIDER, 1972: On a problem of C. Rényi concerning Julia lines. J. Approx. Theory *6*, 312–315. MR *49*, 104. *IV*:5.

H. BEHNKE and F. SOMMER, 1962: Theorie der analytischen Funktionen einer komplexen Veränderlichen (2nd Ed.). Springer, Berlin-Göttingen-Heidelberg. MR *26*, 977. Zbl *101*, 295. *I*:1.

V. I. BELYI, 1965: On the constructive properties of certain classes of functions, continuous in regions with angles (Ukrainian). Dopovidi Akad. Nauk Ukrain. RSR *1965*, 273–276. MR *32*, 475. *I*:6.

______, 1977: Conformal mappings and approximation of analytic functions in domains with quasiconformal boundary (Russian). Mat. Sb. (N. S.) *102* (*144*), 331–361. Translation: Math. USSR-Sb. *31* (1977), 289–317. MR *57*, 91. *I*:6.

V. I. BELYI and V. M. MIKLJUKOV, 1974: Certain properties of conformal and quasiconformal mappings, and direct theorems of the constructive theory of functions (Russian). Izv. Akad. Nauk SSSR Ser. Mat. *38*, 1343–1361. Translation: Math. USSR-Izv. *8* (1974), 1323–1341. MR *52*, 1202. *I*:6.

S. BERGMAN, 1970: The kernel function and conformal mapping (2nd edition). Mathematical Surveys, No. V. American Mathematical Society, Providence, R.I. MR *58*, 3364. *I*:1.

D. L. BERMAN, 1971: On the S. N. Bernstein — I. Marcinkiewicz method for the summability of interpolation processes (Russian). Izv. Vysš. Učebn. Zaved. Matematika *1971*, no. 5, 14–17. MR *45*, 1299. *II*:4.

L. BIEBERBACH, 1934: Lehrbuch der Funktionentheorie, Band I (4th edition). Teubner, Leipzig. Zbl *11*, 358. *IV*:5.

L. BIJVOETS, W. HOGEVEEN, and J. KOREVAAR, 1980: Inverse approximation theorems of Lebedev and Tamrazov. Preprint, University of Amsterdam. *I*:6.

E. BISHOP, 1960: Boundary measures of analytic differentials. Duke Math. J. *27*, 331–340. MR *22*, 1630. *III*:2.

R. P. BOAS, 1954: Entire functions. Academic Press, New York. MR *16*, 914. *IV*:4.

J. E. BRENNAN, 1973: Invariant subspaces and weighted polynomial approximation. Ark. Mat. *11*, 167–189. MR *50*, 403. *I*:3.

———, 1977: Approximation in the mean by polynomials on non-Carathéodory domains. Ark. Mat. *15*, 117–168. MR *56*, 1199. *I*:3.

———, 1980. Point evaluations, approximation in the mean and analytic continuation. Proc. Complex Approximation, Birkhäuser, 32–46. MR *82d*, 1459. *I*:3.

L. BROWN and P. M. GAUTHIER, 1973: The local range set of a meromorphic function. Proc. Amer. Math. Soc. *41*, 518–524. MR *48*, 756. *IV*:3.

L. BROWN, P. M. GAUTHIER, and W. SEIDEL, 1974: Complex approximation for vector-valued functions with an application to boundary behaviour. Trans. Amer. Math. Soc. *191*, 149–163. MR *49*, 1373. *IV*:2.

———, 1975: Possibility of complex approximation on closed sets. Math. Ann. *218*, 1–8. MR *54*, 1504. *IV*:3,5.

I. N. BRUI, 1976: The approximation of functions by generalized means of their expansions in Faber polynomials in domains with corners. Vesci Akad. Navuk BSSR Ser. Fiz.-Mat. Navuk *1976*, no. 1, 14–20, 138–139. MR *57*, 2192. *I*:6.

J. BURBEA, 1976: Polynomial density in Bers spaces. Proc. Amer. Math. Soc. *62*, 89–94. MR *54*, 1851. *I*:3.

———, 1977: Polynomial approximation in Bers spaces of Carathéodory domains. J. London Math. Soc. (2) *15*, 255–266. MR *56*, 92. *I*:3.

______, 1978: Polynomial approximation in Bers spaces of non-Carathéodory domains. Ark. Mat. *16*, 229–234. Zbl *409*, 136. MR *83b*, 551. *I*:3.

B. E. CAIN, 1974: Every direction a Julia direction. Proc. Amer. Math. Soc. *46*, 250–252. MR *50*, 350. *IV*:5.

T. CARLEMAN, 1927: Sur un théorème de Weierstrass. Ark. Mat. Astronom. Fys. *20B*, 1–5. JB *53*, 237. *IV*:3.

L. CARLESON, 1964: Mergelyan's theorem on uniform polynomial approximation. Math. Scand. *15*, 167–175. MR *33*, 1077. *III*:2.

A. S. CAVARETTA, A. SHARMA, and R. S. VARGA, 1980: Interpolation in the roots of unity: An extension of a theorem of J. L. Walsh. Resultate Math. *3*, 155–191. MR *82j*, 4243. *II*:4.

C. K. CHUI and M. N. PARNES, 1971: Approximation by overconvergence of a power series. J. Math. Anal. Appl. *36*, 693–696. MR *45*, 100. *III*:2.

J. H. CURTISS, 1935: Interpolation in regularly distributed points. Trans. Amer. Math. Soc. *38*, 458–473. Zbl *13*, 107. *II*:2,4.

______, 1941a: Necessary conditions in the theory of interpolation in the complex domain. Ann. of Math. (2) *42*, 634–646. MR *3*, 115. *II*:2.

______, 1941b: Riemann sums and the fundamental polynomials of Lagrange interpolation. Duke Math. J. *8*, 525–532. MR *3*, 115. *II*:2.

______, 1962: Interpolation by harmonic polynomials. SIAM J. Appl. Math. *10*, 709–736. MR *28*, 804. *II*:2.

______, 1964: Harmonic interpolation in Fejér points with the Faber polynomials as a basis. Math. Z. *86*, 75–92. MR *29*, 1128. *II*:2.

______, 1966: Solutions of the Dirichlet problem in the plane by approximation with Faber polynomials. SIAM J. Numer. Anal. *3*, 204–228. MR *34*, 1441. *II*:2.

______, 1969a: Transfinite diameter and harmonic polynomial interpolation. J. Analyse Math. *22*, 371–389. MR *40*, 59. *II*:2.

______, 1969b: The asymptotic value of a singular integral related to the Cauchy-Hermite interpolation formula. Aequationes Math. *3*, 130–148. MR *40*, 1072. *II*:2.

______, 1971: Faber polynomials and the Faber series. Amer. Math. Monthly *78*, 577–596. MR *45*, 400. *I*:6.

P. J. DAVIS, *1963: Interpolation and approximation. Blaisdell, New York-Toronto-London (Reprint: Dover, 1975). MR *28*, 82; MR *52*, 154. *I*:2; *II*:1.

______, 1969: Additional simple quadratures in the complex plane. Aequationes Math. *3*, 149–155. MR *40*, 578. *I*:5.

B. L. DINCEN, 1964: The deviation of analytic functions from the mean arithmetic partial sums of the Faber series (Russian). Dokl. Akad. Nauk SSSR *157*, 250–253. Translation: Soviet Math. Dokl. *5* (1964), 909–912. MR *29*, 698. *I*:6.

M. DIXON and J. KOREVAAR, 1977: Approximation by lacunary polynomials. Nederl. Akad. Wetensch. Proc. Ser. A *80* = Indag. Math. *39*, 176–194. MR *56*, 2074. *III*:2.

D. DRASIN, 1974a: A meromorphic function with assigned Nevanlinna deficiencies. Proc. Symp. Complex Analysis Canterbury 1973, 31–41. London Math. Soc. Lecture Note Ser., No. 12. Cambridge University Press, London 1974. MR *53*, 1551. *IV*:5.

______, 1974b: A meromorphic function with assigned Nevanlinna deficiencies. Bull. Amer. Math. Soc. *80*, 766–768. MR *49*, 1697. *IV*:5.

D. DRASIN and A. WEITSMAN, 1976: On the Julia directions and Borel directions of entire functions. Proc. London Math. Soc. (3) *32*, 199–212. MR *53*, 466. *IV*:5.

P. L. DUREN, 1970: Theory of H^p spaces. Academic Press, New York-London. MR *42*, 640. *I*:6.

E. M. DYNKIN, 1974: Uniform polynomial approximation in the complex domain (Russian). Zap. Naučn. Sem. Leningrad. Otdel. Mat. Inst. Steklov. (LOMI) *47*, 164–165. Translation: J. Soviet Math. *9* (1978), 269–271. MR *52*, 503. *I*:6.

______, 1977: Uniform approximation of functions in Jordan domains (Russian). Sibirsk. Mat. Ž. *18*, 775–786, 956. Translation: Siberian Math. J. *7* (1977), 548–557. MR *56*, 1633. *I*:6.

______, 1980: Smoothness of Cauchy type integrals (Russian). Dokl. Akad. Nauk SSSR *250*, 794–797. Translation: Soviet Math. Dokl. *21* (1980), 199–202. MR *81c*, 920. *I*:6.

V. K. DZJADYK, 1962: On the approximation of continuous functions in closed regions with corners and on a problem of S. M. Nikol'skiǐ. I (Russian). Izv. Akad. Nauk SSSR Ser. Mat. *26*, 797–824. Translation: Translations Amer. Math. Soc. *53* (1966), 221–252. MR *27*, 66. *I*:6.

______, 1963a: On the theory of approximation of continuous functions in closed regions and on a problem of S. M. Nikol'skiǐ. II (Russian). Izv. Akad. Nauk SSSR Ser. Mat. *27*, 1135–1164. Translation: Translations Amer. Math. Soc. *53* (1966), 253–284. MR *27*, 938. *I*:6.

______, 1963b: Inverse theorems in the theory of approximation of functions in complex domains (Russian). Ukrain. Mat. Ž. *15*, 365–375. MR *35*, 810; Zbl *119*, 290. *I*:6.

______, 1966: The approximation of analytic functions in regions with smooth and piecewise smooth boundary (Russian). Third Math. Summer School, Kaciveli 1965. Izdat. Naukova Dumka, Kiev. MR *37*, 303. *I*:6.

______, 1968: The constructive properties of functions of Hölder classes on closed sets with piecewise smooth boundary admitting zero angles (Russian). Ukrain. Mat. Ž. *20*, 603–619. Translation: Ukrainian Math. J. *20* (1968), 523–535. MR *38*, 620. *I*:6.

______, 1969: Investigations in the theory of approximations of analytic functions conducted at the Institute of Mathematics of the Ukrainian Academy of Sciences (Russian). Ukrain. Mat. Ž. *21*, 173–192. Translation: Ukrainian Math. J. *21* (1969), 143–159. MR *39*, 543. *I*:6.

______, 1972: The application of generalized Faber polynomials to the approximation of integrals of Cauchy type and functions of the classes A^r in domains with smooth and piecewise smooth boundary (Russian). Ukrain. Mat. Ž. *24*, 3–19. Translation: Ukrainian Math. J. *24* (1972), 1–13. MR *47*, 647. *I*:6.

______, 1975: On the theory of the approximation of functions on closed sets of the complex plane (apropos of a certain problem of S. M. Nikol'skiĭ) (Russian). Trudy Mat. Inst. Steklov. *134*, 63–114, 408. Translation: Proc. Steklov Inst. Math. *134* (1975), 75–130. MR *53*, 831. *I*:6.

______, *1977: Introduction to the theory of uniform approximation of functions by polynomials (Russian). Moscow. MR *58*, 4351. *I*:6.

V. K. DZJADYK and G. A. ALIBEKOV, 1968: Uniform approximation of functions of a complex variable on closed sets with corners (Russian). Mat. Sb. (N. S.) *75 (117)*, 502–557. Translation: Math. USSR-Sb. *4* (1968), 463–517. MR *37*, 81. *I*:6.

V. K. DZJADYK and D. M. GALAN, 1965: The approximation of analytic functions in domains with smooth boundary (Russian). Ukrain. Mat. Ž. *17*, 26–38. MR *33*, 522. *I*:6.

V. K. DZJADYK and A. I. ŠVAI, 1971: The approximation of functions of Hölder classes on closed sets with sharp exterior angles (Russian). Metric questions of the theory of functions and mappings, 74–164; Izdat. Naukova Dumka, Kiev. MR *45*, 987. *I*:6.

W. EIDEL, 1979: Konforme Abbildung mehrfach zusammenhängender Gebiete durch Lösung von Variationsproblemen. Diplomarbeit, Giessen. *I*:5.

B. EPSTEIN, *1965: Orthogonal families of analytic functions. Macmillan, New York. MR *31*, 891. *I*:1.

G. FABER, 1903: Über polynomische Entwicklungen. Math. Ann. *57*, 398–408. JB *34*, 430. *I*:6.

O. J. FARRELL, 1932: On approximation to a mapping function by polynomials. Amer. J. Math. *54*, 571–578. Zbl *4*, 403. *I*:3.

______, 1934: On approximation to an analytic function by polynomials. Bull. Amer. Math. Soc. *40*, 908–914. Zbl *10*, 348. *I*:3.

______, 1966: On approximation measured by a surface integral. SIAM J. Numer. Anal. *3*, 236–247. MR *34*, 318. *I*:3.

L. FEJÉR, 1916: Über Interpolation. Göttinger Nachr. *1916*, 66–91. JB *46*, 419. *II*:4.

______, 1918: Interpolation und konforme Abbildung. Göttinger Nachr. *1918*, 319–331. JB *46*, 517. *II*:2.

M. FEKETE, 1926: Über Interpolation. Z. Angew. Math. Mech. *6*, 410–413. JB *52*, 302. *II*:2.

W. H. J. FUCHS, *1968: Théorie de l'approximation des fonctions d'une variable complexe. Université de Montréal. MR *41*, 1037. *IV*:2,4,5.

W. H. J. FUCHS and W. K. HAYMAN, 1962: An entire function with assigned deficiencies. Studies in mathematical analysis and related topics, 117–125. Stanford Univ. Press, Stanford, Calif. MR *27*, 736. *IV*:5.

D. GAIER, 1954: Über Interpolation in regelmässig verteilten Punkten mit Nebenbedingungen. Math. Z. *61*, 119–133. MR *16*, 812. *II*:2,4.

———, 1964: Konstruktive Methoden der konformen Abbildung. Springer, Berlin-Göttingen-Heidelberg. MR *33*, 1291; Zbl *132*, 367. *I*:1,5.

———, 1977: Approximation durch Fejér-Mittel in der Klasse A. Mitt. Math. Sem. Giessen *123*, 1–6. MR *56*, 1633; Zbl *358*, 161. *I*:6.

———, 1983: Remarks on Alice Roth's fusion lemma. J. Approx. Theory *37*, 246–250. MR *84h*, 3175. *III*:4.

T. W. GAMELIN, *1969: Uniform algebras. Prentice-Hall, Englewood Cliffs, N. J. MR *53*, 1973. *III*:3.

T. H. GANELIUS, 1973: Degree of approximation by polynomials on compact plane sets. Proc. Internat. Sympos., Univ. Texas, Austin, 347–351. Academic Press, New York. MR *49*, 579. *I*:6.

J. GARNETT, 1968: On a theorem of Mergelyan. Pacific J. Math. *26*, 461–467. MR *38*, 287. *III*:3.

P. GAUTHIER, 1969: Tangential approximation by entire functions and functions holomorphic in a disc. Izv. Akad. Nauk Armjan. SSR Ser. Mat. *4*, 319–326. MR *43*, 1172. *IV*:3.

P. M. GAUTHIER, M. GOLDSTEIN, and W. H. OW, 1980: Uniform approximation on unbounded sets by harmonic functions with logarithmic singularities. Trans. Amer. Math. Soc. *261*, 169–183. MR *81m*, 4949. *III*:4.

P. M. GAUTHIER and W. HENGARTNER, 1977: Complex approximation and simultaneous interpolation on closed sets. Canad. J. Math. *29*, 701–706. MR *58*, 4214. *IV*:3.

P. GAUTHIER and W. SEIDEL, 1971: Some applications of Arakelian's approximation theorems to the theory of cluster sets. Izv. Akad. Nauk Armjan. SSR Ser. Mat. *6*, 458–464. MR *46*, 350. *IV*:5.

A. H. GERMAN, 1980: Interpolation in the complex domain. Anal. Math. *6*, 121–135. MR *81m*, 4949. *II*:4.

I. GLICKSBERG and J. WERMER, 1963: Measures orthogonal to a Dirichlet algebra. Duke Math. J. *30*, 661–666. MR *27*, 1176. *III*:2.

G. M. GOLUZIN [GOLUSIN], 1969: Geometric theory of functions of a complex variable. Translations of Mathematical Monographs, Vol. 26. American Mathematical Society, Providence, R. I. MR *40*, 56. *I*:3; *II*:3.

T. H. GRONWALL, 1920: A sequence of polynomials connected with the n-th roots of unity. Bull. Amer. Math. Soc. *27*, 275–279. JB *48*, 395. *II*:4.

F. HARTOGS and A. ROSENTHAL, 1931: Über Folgen analytischer Funktionen. Math. Ann. *104*, 606–610. Zbl *1*, 213. *III*:3.

V. P. HAVIN, 1968a: Polynomial approximation in the mean in certain non-Carathéodory regions. I. (Russian). Izv. Vysš. Učebn. Zaved. Matematika *1968*, no. 9 (76), 86–93. MR *38*, 1090. *I*:3.

______, 1968b: Polynomial approximation in the mean in certain non-Carathéodory regions. II. (Russian). Izv. Vysš. Učebn. Zaved. Matematika *1968*, no. 10 (77), 87–94. MR *40*, 69. *I*:3.

W. K.. HAYMAN, 1964: Meromorphic functions. Clarendon Press, Oxford. MR *29*, 263. *IV*:5.

L. I. HEDBERG, 1965: Weighted mean square approximation in plane regions, and generators of an algebra of analytic functions. Ark. Mat. *5*, 541–552. MR *36*, 572. *I*:3.

______, 1969: Weighted mean approximation in Carathéodory regions. Math. Scand. *23*, 113–122. MR *41*, 372. *I*:3.

E. HLAWKA, 1969: Interpolation analytischer Funktionen auf dem Einheitskreis. Number Theory and Analysis (Papers in Honor of Edmund Landau), 97–118. Plenum, New York. MR *42*, 1135. *II*:4.

L. HOISCHEN, 1967: A note on the approximation of continuous functions by integral functions. J. London Math. Soc. *42*, 351–354. MR *35*, 385. *IV*:3.

______, 1970: Asymptotische Approximation stetiger Funktionen durch ganze Dirichlet-Reihen. J. Approx. Theory *3*, 293–299. MR *41*, 1635. *IV*:3.

______, 1975: Approximation und Interpolation durch ganze Funktionen. J. Approx. Theory *15*, 116–123. MR *52*, 2012. *IV*:3.

W. KAPLAN, 1955: Approximation by entire functions. Michigan Math. J. *3*, 43–52. MR *17*, 31. *IV*:3,5.

E. M. KEGEJAN, 1966: Boundary behavior of unbounded analytic functions defined in a disc (Russian). Akad. Nauk Armjan. SSR Dokl. *42*, 65–72. MR *35*, 1040. *IV*:5.

M. KELDYSH [KELDYCH], 1939: Sur l'approximation en moyenne quadratique des fonctions analytiques. Mat. Sbornik N. S. *5* (*47*), 391–401. MR *2*, 80. *I*:3.

______, 1945a: Sur l'approximation des fonctions holomorphes par les fonctions entieres. C. R. (Doklady) Acad. Sci. URSS (N. S.) *47*, 239–241. MR *7*, 150. *IV*:2,4.

______, 1945b: Sur la représentation par des séries de polynomes des fonctions d'une variable complexe dans de domaines fermés. (Russian). Mat. Sbornik N. S. *16* (*58*), 249–258. MR *7*, 285. *III*:2.

M. KELDYSH and M. LAVRENTIEV [KELDYCH and LAVRENTIEFF], 1939: Sur une problème de M. Carleman. C. R. (Doklady) Acad. Sci. URSS (N. S.) *23*, 746–748. MR *2*, 82; JB *65*, 1227. *IV*:2,3.

L. I. KOLESNIK and M. I. ANDRAŠKO, 1971: Inverse theorems on approximation in the mean in domains with angles (Russian). Ukrain. Mat. Ž. *23*, 97–104. Translation: Ukrainian Math. J. *23* (1971), 85–91. MR *44*, 91. *I*:6.

J. KOREVAAR, *1980: Polynomial and rational approximation in the complex domain. Aspects of contemporary complex analysis. Proc. Conf., Univ. of Durham, 1979, 251–292. Academic Press, New York. MR *82j*, 4244.

J. KOREVAAR and M. DIXON, 1978: Lacunary polynomial approximation. Linear spaces and approximation (Proc. Conf., Math. Res. Inst., Oberwolfach, 1977), 479–489. Internat. Ser. Numer. Math., Vol. 40. Birkhäuser. MR *58*, 2568. *III*:2.

T. KÖVARI, 1968: On the uniform approximation of analytic functions by means of interpolation polynomials. Comment. Math. Helv. *43*, 212–216. MR *37*, 1007. *II*:2.

______, 1971: On the distribution of Fekete points. II. Mathematika *18*, 40–49. MR *44*, 1264. *II*:2.

______, 1972: On the order of polynomial approximation for closed Jordan domains. J. Approx. Theory *5*, 362–373. MR *49*, 108. *I*:6.

T. KÖVARI and CH. POMMERENKE, 1967: On Faber polynomials and Faber expansions. Math. Z. *99*, 193–206. MR *37*, 558. *I*:6.

I. V. KULIKOV, 1979: L_p-convergence of Bieberbach polynomials (Russian). Izv. Akad. Nauk SSSR Ser. Mat. *43*, 1121–1144. Translation: Math. USSR-Izv. *15* (1980), 349–371. MR *81a*, 99. *I*:5.

M. A. LAVRENTIEV [LAVRENTIEFF], 1936: Sur les fonctions d'une variable complexe représentables par des séries de polynomes. Hermann, Paris. JB *62*, 1205. *IV*:3.

N. A. LEBEDEV and N. A. ŠIROKOV, 1971: The uniform approximation of functions on closed sets that have a finite number of angular points with nonzero exterior angles (Russian). Izv. Akad. Nauk Armjan. SSR Ser. Mat. *6*, 311–341. MR *45*, 987. *I*:6.

N. A. LEBEDEV and P. M. TAMRAZOV, 1970: Inverse approximation theorems on regular compacta of the complex plane (Russian). Izv. Akad. Nauk SSSR Ser. Mat. *34*, 1340–1390. Translation: Math. USSR-Izv. *4* (1970), 1355–1405. MR *45*, 436. *I*:6.

O. LEHTO, 1955: On the first boundary value problem for functions harmonic in the unit circle. Ann. Acad. Sci. Fenn. Ser. A. I. no. *210*, 26 pp. MR *17*, 960. *IV*:5.

O. LEHTO and K. I. VIRTANEN, 1965: Quasikonforme Abbildungen. Springer-Verlag, Berlin-New York. MR *32*, 995. *I*:6.

F. LEJA, 1957: Sur certaines suites liées aux ensembles plans et leur application à la représentation conforme. Ann. Polon. Math. *4*, 8–13. MR *20*, 1171. *II*:2.

F. D. LESLEY, V. S. VINGE, and S. E. WARSCHAWSKI, 1974: Approximation by Faber polynomials for a class of Jordan domains. Math. Z. *138*, 225–237. MR *50*, 687. *I*:6.

D. LEVIN, N. PAPAMICHAEL, and A. SIDERIDIS, 1978: The Bergman kernel method for the numerical conformal mapping of simply connected domains. J. Inst. Math. Appl. *22*, 171–187. MR *80h*, 3000. *I*:5.

K. LÖWNER, 1919: Über Extremumsätze bei der konformen Abbildung des Äusseren des Einheitskreises. Math. Z. *3*, 65–77. JB *47*, 325. *I*:6.

S. LOSINSKY, 1939: Sur le procédé d'interpolation de Fejér. C. R. (Doklady) Akad. Sci. URSS (N. S.) *24*, 318-321. MR *1*, 333. *II*:4.

W. LUH, 1974: Über die Summierbarkeit der geometrischen Reihe. Mitt. Math. Sem. Giessen *113*. MR *56*, 130. *III*:2.

______, 1976: Über den Satz von Mergelyan. J. Approx. Theory *16*, 194–198. MR *55*, 91. *III*:2.

J. MARCINKIEWICZ, 1936: Sur l'interpolation (I). Studia Math. *6*, 1–17. Zbl *16*, 19. *II*:4.

A. I. MARKUSHEVICH [MARKUSCHEWITSCH], 1934: Conformal mapping of regions with variable boundary and application to the approximation of analytic functions by polynomials (Russian). Dissertation. Moscow. *I*:3.

M. S. MELNIKOV and S. O. SINANYAN, *1976: Aspects of approximation theory for functions of one complex variable. J. Soviet Math. *5*, 688–752. MR *58*, 2569. *I*:3; *III*:3.

K. MENKE, 1972: Extremalpunkte und konforme Abbildung. Math. Ann. *195*, 292–308. MR *45*, 92. *II*:2.

______, 1974a: Zur Approximation des transfiniten Durchmessers bei bis auf Ecken analytischen geschlossenen Jordankurven. Israel J. Math. *17*, 136–141. MR *50*, 1032. *II*:2.

______, 1974b: Bestimmung von Näherungen für die konforme Abbildung mit Hilfe von stationären Punktsystemen. Numer. Math. *22*, 111–117. MR *51*, 831. *II*:2.

______, 1975: Lösung des Dirichlet-Problems bei Jordangebieten mit analytischem Rand durch Interpolation. Monatsh. Math. *80*, 297–306. MR *52*, 1575. *II*:2.

______, 1976: Über die Verteilung von gewissen Punktsystemen mit Extremaleigenschaften. J. Reine Angew. Math. *283/284*, 421–435. MR *53*, 1180. *II*:2.

______, 1977: Über das von Curtiss eingeführte Maximalpunktsystem. Math. Nachr. *77*, 301–306. MR *56*, 92. *II*:2.

S. N. MERGELYAN, 1951: On the representation of functions by series of polynomials on closed sets (Russian). Dokl. Akad. Nauk SSSR (N. S.) *78*, 405–408. Translation: Translations Amer. Math. Soc. *3* (1962), 287–293. MR *13*, 23; *14*, 858. *III*:2.

______, *1952: Uniform approximations to functions of a complex variable (Russian). Uspehi Mat. Nauk (N. S.) *7*, no. 2 (48), 31–122. Translation: Translations Amer. Math. Soc. *3* (1962), 294–391. MR *14*, 547; *15*, 612. *I*:6; *III*:2,3; *IV*:2,3,4.

______, *1953: On the completeness of systems of analytic functions (Russian). Uspehi Mat. Nauk (N. S.) *8*, no. 4 (56), 3–63. Translation: Translations Amer. Math. Soc. *19* (1962), 109–166. MR *15*, 411; MR *24A*, 258. *I*:3.

______, 1955: General metric criteria of completeness of a system of polynomials (Russian). Dokl. Akad. Nauk SSSR (N. S.) *105*, 901–904. MR *17*, 730. *I*:3.

______, *1956: Weighted approximations by polynomials (Russian). Uspehi Mat. Nauk (N. S.) *11*, no. 5 (71), 107–152. Translation: Translations Amer. Math. Soc. *10* (1958), 59–106. MR *18*, 734; MR *20*, 190. *I*:3.

I. P. NATANSON, 1955: Konstruktive Funktionentheorie. Akademie-Verlag, Berlin. MR *16*, 1100. *II*:4.

Z. NEHARI, 1952: Conformal mapping. McGraw-Hill, New York (Reprint: Dover, 1975). MR *13*, 640. *I*:1,5.

A. A. NERSESJAN, 1971: Carleman sets (Russian). Izv. Akad. Nauk Armjan. SSR Ser. Mat. *6*, 465–471. MR *46*, 66. *IV*:3.

______, 1972: Uniform and tangential approximation by meromorphic functions (Russian). Izv. Akad. Nauk Armjan. SSR Ser. Mat. *7*, 405–412. MR *51*, 484. *IV*:1,3.

______, 1973: The simultaneous tangential approximation of functions and their derivatives (Russian). Izv. Akad. Nauk Armjan. SSR Ser. Mat. *8*, 464–473. MR *51*, 484. *IV*:3.

______, 1978: A theorem on simultaneous tangential approximation (Russian). Izv. Akad. Nauk Armjan. SSR Ser. Mat. *13*, 442–447. MR *81b*, 500. *IV*:3.

______, 1980: On uniform approximation with simultaneous interpolation by analytic functions (Russian). Izv. Akad. Nauk Armjan. SSR Ser. Mat. *15*, 249–257. MR *82g*, 2914. *IV*:3.

M. H. A. NEWMAN, 1951: Elements of the topology of plane sets of points. University Press, Cambridge. MR *13*, 483. *IV*:2.

V. PAATERO, 1933: Über Gebiete von beschränkter Randdrehung. Ann. Acad. Sci. Fenn. *A 37*, No. 9. Zbl *6*, 354. *I*:6.

CH. POMMERENKE, 1964: Über die Faberschen Polynome schlichter Funktionen. Math. Z. *85*, 197–208. MR *29*, 1129. *I*:6.

———, 1965: Konforme Abbildung und Fekete-Punkte. Math. Z. *89*, 422–438. MR *31*, 1073. *I*:6.

———, 1967: Über die Verteilung der Fekete-Punkte. Math. Ann. *168*, 111–127. MR *34*, 1105. *II*:2.

———, 1969: Über die Verteilung der Fekete-Punkte. II. Math. Ann. *179*, 212–218. MR *40*, 70. *II*:2.

———, 1975: Univalent functions. Vandenhoeck-Ruprecht, Göttingen. MR *58*, 3367. *I*:6; *II*:2; *III*:2.

I. I. PRIVALOV [PRIWALOW], 1956: Randeigenschaften analytischer Funktionen. Deutscher Verlag der Wissenschaften, Berlin. MR *18*, 727; Zbl *45*, 347; Zbl *73*, 65. *I*:6; *IV*:5.

J. RADON, 1919: Über die Randwertaufgaben beim logarithmischen Potential. Sitz.-Ber. Wien. Akad. Wiss., Abt. IIa, *128*, 1123–1167. JB *47*, 457. *I*:6.

A. ROTH, 1938: Approximationseigenschaften und Strahlengrenzwerte meromorpher und ganzer Funktionen. Comment. Math. Helv. *11*, 77–125. Zbl *20*, 235. *III*:3; *IV*:1,2,3,5.

———, 1973: Meromorphe Approximationen. Comment. Math. Helv. *48*, 151–176. MR *57*, 2193; Zbl *275*, 202. *IV*:1,2,3.

———, 1976: Uniform and tangential approximations by meromorphic functions on closed sets. Canad. J. Math. *28*, 104–111. MR *57*, 1305. *III*:4; *IV*:1,2,3.

L. A. RUBEL and S. VENKATESWARAN, 1976: Simultaneous approximation and interpolation by entire functions. Arch. Math. *27*, 526–529. MR *55*, 813. *IV*:3.

W. RUDIN, 1956: Subalgebras of spaces of continuous functions. Proc. Amer. Math. Soc. *7*, 825–830. MR *18*, 587. *III*:2.

———, 1974: Real and complex analysis (2nd edition). McGraw-Hill, New York-Düsseldorf-Johannesburg. MR *49*, 1616. *III*:2.

C. RUNGE, 1885: Zur Theorie der eindeutigen analytischen Funktionen. Acta Math. *6*, 228–244. JB *17*, 379. *II*:3; *III*:1.

S. SCHEINBERG, 1976: Uniform approximation by entire functions. J. Analyse Math. *29*, 16–18. MR *58*, 3388. *IV*:3.

W. J. SCHNEIDER, 1980: Approximation and harmonic measure. Aspects of contemporary complex analysis. Proc. Conf., Univ. of Durham, 1979, 333–349. Academic Press, New York. MR *82g*, 2913. *IV*:5.

I. A. ŠEVČUK, 1973: Constructive characterization of functions of the classes $D^rH\omega_2(t)$ on closed sets with a piece-wise smooth boundary (Russian). Ukrain. Mat. Ž. *25*, 81–90, 142. Translation: Ukrainian Math. J. *25* (1973), 66–73. MR *47*, 957. *I*:6.

W. E. SEWELL, *1942: Degree of approximation by polynomials in the complex domain. Princeton. MR *4*, 78. *I*:6.

H. S. SHAPIRO, 1967: Some observations concerning weighted polynomial approximation of holomorphic functions (Russian). Mat. Sb. (N. S.) *73* (*115*), 320–330. Translation: Math. USSR-Sb. *2* (1967), 285–294. MR *36*, 89. *I*:3.

Y. C. SHEN, 1936: On interpolation and approximation by rational functions with preassigned poles. J. Chinese Math. Soc. *1*, 154–173. Zbl *15*, 251. *II*:3.

J. SICIAK, 1965: Some applications of interpolating harmonic polynomials. J. Analyse Math. *14*, 393–407. MR *31*, 432. *II*:2.

I. B. SIMONENKO, 1978: Convergence of Bieberbach polynomials in the case of a Lipschitz domain (Russian). Izv. Akad. Nauk SSSR Ser. Mat. *42*, 870–878. Translation: Math. USSR-Izv. *13* (1979), 166–174. MR *81m*, 4946. *I*:5.

S. O. SINANJAN, 1966: Approximation by analytic functions and polynomials in the mean with respect to the area (Russian). Mat. Sb. (N. S.) *69* (*111*), 546–578. Translation: Translations Amer. Math. Soc. *74* (1968), 91–124. MR *35*, 78. *I*:3.

A. SINCLAIR, 1965: $|\epsilon(z)|$-closeness of approximation. Pacific J. Math. *15*, 1405–1413. MR *32*, 1328. *III*:3.

N. A. ŠIROKOV, 1971: The uniform approximation of functions on closed sets without angular points (Russian). Zap. Naučn. Sem. Leningrad. Otdel. Mat. Inst. Steklov. (LOMI) *22*, 209–211. Translation: J. Soviet Math. *2* (1974), 235–237. MR *45*, 1299. *I*:6.

______, 1972: Uniform approximation of functions on closed sets that have a finite number of angular points with nonzero exterior angles (Russian). Dokl. Akad. Nauk SSSR *205*, 798–800. Translation: Soviet Math. Dokl. *13* (1972), 1041–1044. MR *46*, 1622. *I*:6.

______, 1974a: The uniform approximation of functions on closed sets with non-zero exterior angles (Russian). Izv. Akad. Nauk Armjan. SSR Ser. Mat. *9*, 62–80, 83. MR *50*, 352. *I*:6.

______, 1974b: Weighted approximations on closed sets with corners (Russian). Dokl. Akad. Nauk SSSR *214*, 295–297. Translation: Soviet Math. Dokl. *15* (1974), 143–146. MR *51*, 484. *I*:6.

______, 1976: Approximation of continuous analytic functions in domains with bounded boundary rotation (Russian). Dokl. Akad. Nauk SSSR *228*, 809–812. Translation: Soviet Math. Dokl. *17* (1976), 844–847. MR *54*, 88. *I*:6.

V. I. SMIRNOV and N. A. LEBEDEV, *1968: Functions of a complex variable: Constructive theory. The M.I.T. Press, Cambridge, Mass. MR *37*, 1000. *I*:3,6; *II*:1.

A. STRAY, 1974: Characterization of Mergelyan sets. Proc. Amer. Math. Soc. *44*, 347–352. MR *50*, 1860. *IV*:2.

______, 1977a: Approximation by analytic functions which are uniformly continuous on a subset of their domain of definition. Amer. J. Math. *99*, 787–800. MR *58*, 3374. *IV*:2.

______, 1977b: On a theorem of Arakelian. Manuscript. *IV*:2.

______, 1978: On uniform and asymptotic approximation. Math. Ann. *234*, 61–68. MR *57*, 1305. *IV*:2,5.

______, 1980: Vitushkin's method and some problems in approximation theory. Aspects of contemporary complex analysis. Proc. Conf., Univ. of Durham, 1979, 351–365. Academic Press, New York. MR *82g*, 2915. *IV*:2.

P. K. SUETIN, *1964: The basic properties of Faber polynomials (Russian). Uspehi Mat. Nauk *19*, 125–154. Translation: Russian Math. Surveys *19* (1964), 121–149. MR *29*, 1129. *I*:6.

______, 1969: Areally orthogonal polynomials and Bieberbach polynomials (Russian). Dokl. Akad. Nauk SSSR *188*, 294–296. Translation: Soviet Math. Dokl. *10* (1969), 1123–1126. MR *45*, 1609. *I*:2.

______, *1971: Polynomials orthogonal over a region and Bieberbach polynomials (Russian). Trudy Mat. Inst. Steklov. *100*. Translation: Proc. Steklov Inst. Math. *100*. Amer. Math. Soc., 1974. MR *57*, 491. *I*:2,4,5.

______, 1972: Certain asymptotic and approximation properties of polynomials that are orthogonal over a smooth contour (Russian). Collection of articles, Kalinin. Gos. Univ., Kalinin. MR *52*, 504. *I*:2.

______, *1976: Series in Faber polynomials and several generalizations. J. Soviet Math. *5*, 502–551. MR *56*, 2069. *I*:6.

A. I. ŠVAI, 1973: Approximation of analytic functions by de la Vallée-Poussin polynomials (Russian). Ukrain. Math. Ž. *25*, 848–853, 864. Translation: Ukrainian Math. J. *25* (1973), 710–713. MR *50*, 688. *I*:6.

P. SZŮSZ, 1974: Remark on a theorem of Aharonov and Walsh. Israel J. Math. *17*, 108–110. MR *51*, 484. *I*:4.

P. M. TAMRAZOV, 1971: Solid inverse theorems of polynomial approximation for regular compacta of the complex plane (Russian). Dokl. Akad. Nauk SSSR *198*, 540–542. Translation: Soviet Math. Dokl. *12* (1971), 855–858. MR *44*, 783. *I*:6.

______, 1973: A solid inverse problem of polynomial approximation of functions on a regular compactum (Russian). Izv. Akad. Nauk SSSR Ser. Mat. *37*, 148–164. Translation: Math. USSR-Izv. *7* (1973), 145–162. MR *51*, 1863. *I*:6.

A. E. TAYLOR, 1958: Introduction to functional analysis. Wiley, New York-London-Sydney. MR *20*, 897. *IV*:1.

L. A. TER-ISRAELJAN, 1971: Uniform and tangential approximations of functions that are holomorphic in an angle by meromorphic functions, with an estimate of their growth (Russian). Izv. Akad. Nauk Armjan. SSR Ser. Mat. *6*, 67–80. MR *44*, 783. *IV*:4.

A. F. TIMAN, 1963: Theory of approximation of functions of a real variable. Pergamon Press, Oxford-London-New York-Paris. MR *33*, 82. *I*:6.

P. VÉRTESI, 1980: Linear operators on the roots of unity. Studia Sci. Math. Hung. *15*, 241–245. MR *84e*, 1799. *II*:4.

______, 1983: On discrete linear operators. Studia Sci. Math. Hung. *18*, 429–433. *II*:4.

A. G. VITUSHKIN, 1959: Necessary and sufficient conditions a set should satisfy in order that any function continuous on it can be approximated uniformly by analytic or rational functions (Russian). Dokl. Akad. Nauk SSSR *128*, 17–20. MR *22*, 128. *III*:3.

______, 1966: Conditions on a set which are necessary and sufficient in order that any continuous function, analytic at its interior points, admit uniform approximation by rational fractions (Russian). Dokl. Akad. Nauk SSSR *171*, 1255–1258. Translation: Soviet Math. Dokl. *7* (1966), 1622–1625. MR *35*, 79. *III*:3.

J. L. WALSH, 1933: Note on polynomial interpolation to analytic functions. Proc. Nat. Acad. Sci. U.S.A. *19*, 959–963. Zbl *8*, 19. *II*:2.

______, *1969: Interpolation and approximation by rational functions in the complex domain (5th edition). Amer. Math. Society, Providence, R.I. MR *36*, 349. *I*:1,4; *II*:1,3; *III*:2.

J. L. WALSH and H. G. RUSSELL, 1934: On the convergence and overconvergence of sequences of polynomials of best simultaneous approximation to several functions analytic in distinct regions. Trans. Amer. Math. Soc. *36*, 13–28. Zbl *8*, 214. *II*:3.

S. E. WARSCHAWSKI, 1961: On differentiability at the boundary in conformal mapping. Proc. Amer. Math. Soc. *12*, 614–620. MR *24A*, 253. *I*:6.

H. WITTICH, 1968: Neuere Untersuchungen über eindeutige analytische Funktionen. Springer-Verlag, Berlin-Heidelberg-New York. MR *17*, 1067. *IV*:5.

L. ZALCMAN, *1968: Analytic capacity and rational approximation. Lecture Notes No. 50. Springer-Verlag, Berlin-Heidelberg-New York. MR *37*, 559. *III*:3; *IV*:1.

A. ZYGMUND, 1959: Trigonometric series, Vol. II. University Press, Cambridge. MR *21*, 1208. *II*:4.

INDEX

A

AC-capacity 121
Analytic α-capacity 122
Approximation
  by analytic functions 136ff
  on closed sets 130ff
  of a function and its derivatives 160
  with interpolation 160
  by meromorphic functions 130ff
  by rational functions 86, 109ff
  and uniqueness theorems 171ff
Arakeljan's approximation theorem 139ff, 142
Arithmetic means 88
Asymptotic
  approximation, set of 172
  representation 12
  uniqueness set 172
Augmented basis 36

B

Banach-Steinhaus theorem 87
Barrier reef 92
Bergman kernel function 30
Bernstein's lemma 27, 74
Bessel's inequality 24
Bieberbach polynomials 34ff
Birkhoff-Hermite interpolation 81
Bishop's localization theorem 109, 114, 116ff, 127ff
Boundary behavior in the unit disk 168ff, 174
Boundary behavior of Cauchy integrals 41, 52
Bounded rotation 45
BR 45

C

C_1-means 88
$C(p, \alpha)$ 35
Capacity of ∂G 12, 62
Carathéodory domain 17
Carathéodory set 23
Carleman set 145, 157
Carleman's theorem 145, 149
Cauchy formula 93
Cauchy integral 41, 52
c-condition 55
Cesàro means 88
Chebyshev polynomial 61
Closed subset 16
Cluster set 174
Completeness of an ON system 25
Complete subset 16
CON system 25
Condition (A) 155ff
Condition K 144
Conditions (K_1), (K_2) 139
Constriction function 146
Convergence in the mean 72
Convergence theorem of Kalmar and Walsh 64ff
Convex hull 11

D

Deficiency 177
Dense subset 25
Dirichlet problem for radial limits 170

E

ε-approximation 145, 151ff, 160
Equiconvergence 60, 79

Error function 145
Exceptional values 74
Exhausting sequence 2

F
Faber
  coefficients 44
  domain 49
  expansion 40
  mapping T 47ff
  polynomials 43
  series 44, 52
Fejér means 53
Fejér points 67
Fekete points 70, 73
Fourier
  coefficients 24
  conjugate 41
  series 25
Function spaces $C(K)$, $A(K)$, $P(K)$, $R(K)$ 109, 110
Fusion lemma of Roth 122ff,127

G
Garnett's proof of Bishop's theorem 117
Gauss transform 159
General Cauchy formula 93
Generalized Cauchy integral theorem 98
Generating function for Faber polynomials 44
Gramian determinant 7
Gramian matrix 7
Gram-Schmidt orthogonalization process 6
Green's formula 10, 13
Green's function of a domain 73
Grunsky coefficients 43

H
Harmonic functions, approximation of 129
Hartogs and Rosenthal, theorem of 122
Hermite interpolation 81, 85
Hermite's interpolation formula 58
Hermitian matrix 7
Hilbert space $L^2(G)$ 4

I
Index of multiplicity 177
Inner product 4
Inner snake 17, 20, 92
Interpolating polynomial 58ff
Interpolation 58ff
  error 59
  points 59
Inverse problem of Nevanlinna theory 177

J
Jordan domain 26
Julia direction 168, 175

K
Koebe's ¼ theorem 100

L
Lacunary polynomials 109
Lagrange's formula of interpolation 58, 81, 82
Lebesgue constants 55
Lebesgue measure 2
Legendre polynomials 40
Lemma
  of Bernstein 27, 74
  of Mergelyan 101
  of Roth 122ff
Level curve 73
Linear independence 6
Local connectedness 138
Localization theorem of Bishop 109, 114, 116ff, 127ff

M
Maximal convergence 27, 66, 71, 74
Mean-value property 37
Mergelyan's lemma 101ff
Mergelyan's theorem 97ff

Meromorphic approximation 134ff
m-fold interpolation 59
Minimal norm of polynomials 11
Modulus of continuity of Cauchy integrals 52ff
Moon-shaped domain 20ff
Moving poles of meromorphic functions 95, 137ff

N
Nersesjan's theorem 155ff
Nevanlinna deficiency 177
Node matrix 62
Nodes 63
Non-Hermitian orthogonality relation 40
Noodles 175
n^{th} Faber polynomial 44

O
ON expansions in Hilbert space 24
ON polynomials 11
ON system 6
One-point compactification 133
Orthogonalization process 6
Orthonormal system 6
Outer snake 17, 92

P
PA property 16ff
Parseval's identity 25
Partition of unity 115ff
Pole shifting method 95
Pompeiu formula 99, 125
Positive semidefinite matrix 8

Q
Quality of approximation 53ff
Quasiconformal curve 56

R
Radial limit function 164
Radial limits of entire functions 164ff
Rational approximation 78, 109ff
Reproducing property 30
Roth's approximation theorem 131ff
Roth's fusion lemma 122ff
Runge's approximation theorem 76, 92, 94
Runge's little theorem 96

S
Schneider's noodles 175
Series representation of the Bergman kernel function 31
Set of asymptotic approximation 172
Shifting poles 95, 137, 140
Singularities 36
Slit domain 20
Starlike domain 9
Stitched disk 113, 127
Summability processes 88
Swiss cheese 110ff

T
Tangential approximation 145, 149, 155
Theorem
of Arakeljan 139ff, 142
of Bishop 109, 114, 116ff, 127ff
of Carleman 145, 149
of Hartogs and Rosenthal 122
of Kalmár and Walsh 64ff
of Mergelyan 97ff
of Nersesjan 155ff
of Roth 131ff
of Runge 76, 92, 94
of Vitushkin 109, 121ff
Tietze's extension theorem 99

U
Uniform distribution of nodes 64, 89
Uniqueness and approximation theorems 171ff
Uniqueness set, asymptotic 172
Uniqueness theorems 171ff

V
Vitushkin's theorem 109, 121ff

W
Weierstrass set 142
Weight function 5, 16

Z
Z 53
Zeros of orthogonal polynomials 11, 13
Zygmund class 53